ECHOES ALONG THE DELAWARE

Alice H. Jacobson
Robert Jacobson

Pepacton Press
R.R.1 Box 151, Andes, N.Y. 13731

ISBN 0-9634487-0-6 $18.00

Alice H. Jacobson is author of another book, "Beneath Pepacton Waters", an account of life along the East Branch of the Delaware River and four villages, Pepacton, Shavertown, Union Grove and Arena, as the land was acquired by New York City Board of Water Supply to build the Pepacton Reservoir.

For further information write to:

Pepacton Press

R.R.1 Box 151

Andes, N.Y. 13731

Published by Pepacton Press
R.R. 1 Box 151
Andes, N.Y. 13731

Library of Congress
Catalog Card Number
92-61987

ISBN 0-9634487-0-6

Printed by Reporter Co., Inc.
Walton, N.Y.

To

Our Ancestors who so bravely came to this country

with but a few personal possessions

and

the following generations that continued

to pave the way for

us to have the privilege to live and

explore the beautiful sparkling streams and

green valleys of the Catskill Mountains

and

to future generations

who may choose to rediscover the

paths of past generations.

CONTENTS

ACKNOWLEDGMENTS

We appreciate the time that Mabel and Clara Miller have given us, letting us copy many photos and recording information.

It was a most enjoyable afternoon on April 24, 1990, when David Jacobson, age 92, Isaac Jacobson, age 86, and his wife Dorothy Jacobson gathered at the home of Mabel Miller, age 95, and her daughter, Clara Miller to tell us of their youth during the early 1900's when they lived near Pepacton, N.Y.

Our thanks also go to many other people who have contributed photos and other information; Chip Sidorowicz, Grant March, Delilah Babcock, Betty Berke, Doris Allen, Florence Northrup, Ann Liddle, Aasta Liddle, Wallace Wynkoop, Florence Russell Huggins, Dick and Dora Sprague, Rev. Jacob Bouw, Don and Frances Bramley, Vera Craft, Olive Wilson, Fostor Beach, Dorothy Andrews, Edna Shaver, Dorothy Rice, Jim Andrews, Josephine Huggins Noonan and Clifford Sprague.

Last, but not least, we want to thank our children, John Jacobson and Linda Winchester, for many helpful suggestions.

We have tried to identify all the photos and all the sources of information throughout the book. The photos that are not identified come from our own collection.

INTRODUCTION

Each day we live becomes a day of history. Changes seem to occur gradually and we are so deeply involved with our daily routine and lives that many changes slide by with little recognition and thus many things fade away in time.

We are thankful for our ancestors who so bravely came to this country with but a few personal possessions and settled among the beautiful green valleys and sparkling streams of the Catskill Mountains with a hope of making a better life for themselves and their families. We also appreciate their descendants who continued to pave the way for us to live here among the Catskill Mountains.

It is our intention that this book will record some of the memories of those people. It contains a collection of many stories and old photos. After publishing "Beneath Pepacton Waters", many people wrote of some of their memories of the Pepacton Reservoir area and some people were so kind as to send photos and memorabilia which we have included in this book along with information of the area which surrounds the Pepacton Reservoir.

Roads

Access to Delaware County in the 1700's when the first settlers began to settle in the area as mentioned in "Delaware County, New York History of the Century 1797-1897 Centennial Celebration June 9 and 10, 1897, Edited by David Murray, LL.D., William Clark, Publisher, Delhi, N.Y. 1898," was in three principal directions:

1. Through the Shandaken Mountains from Kingston into Middletown and the valley of the East Branch.
2. By the head-waters of the West Branch through Schoharie and Greene County from Catskill.
3. Up the valley of the Delaware as it winds through the mountains and then up either branch into the various valleys of the county.

As the valley became settled, roads were opened and settlements became connected. In the beginning there were only trails such as the Indians followed on their hunting excursions. Trees had to be cut and rocks had to be removed to make room for horses and wagons to travel through the area. Bridges had to be built over intervening streams. The roads were dusty in summer, muddy in spring and autumn, and only good in winter when they were snow covered. Roads in the smaller valleys were rough and stony and as each shower washed away the earth the stones became more exposed.

Each district elected a pathmaster to care for the roads. Each citizen was assessed for a sum in proportion to the size of his farm. The farmer had a choice of working out his assessment on the roads in hand labor or with a team and driver. This was almost invariably his choice and therefore, the work was not always the best for the wants of the highway. The pathmaster was often ignorant of the correct method of maintaining his district and repairs were often waste labor and roads were left in bad condition.

Turnpike Companies were formed in later years to keep special roads in repair and they were authorized to charge a toll.[1]

Ethel Bussy Coulter in her book of "History of Margaretville and the Surrounding Area" mentions, "The rugged roads from Esopus (Kingston) to Pine Hill were replaced by the Toll Plank Road in 1859.

"The plank roads were made from hemlock after the bark was peeled. This was about the only use made of the hemlock logs after the bark peeling. Most of it was left to decay in the forest where it fell.

"With the heavy traffic of tanning days, the plank road needed yearly repair. Funds were not available for this and the roads were in very bad condition."

The plank road disappeared after the railroad reached Shandaken and later Pine Hill in 1869.

"After the plank road came what was known as the stone tram road. These were used mainly in Ulster County, the great center of the bluestone industry.

"The stone tram road was constructed of two parallel slabs of stone sixteen to eighteen inches wide, one for each wheel of the wagon. They were not too successful, for the large iron-bound wheels and heavy loads soon cut deep channels into the stone, and it was almost impossible to get the wagon out of the deep ruts.

"When the Ashokan Reservoir was being constructed, portions of the old stone tram road were still visable.

"With the collapse of the bluestone industry about 1895, the stone tram road fell into disuse."[2]

"Will Cockburn's map of 1765 shows a road from Marbletown up Espous Creek, over Pine Hill and down the Tweed to Pakatakan, and this was the first road leading into what is now Delaware County. Sauthier's Map of the Province of New York, published Jan. 1, 1779, shows a road laid out from Saugerties to "Pakataghkan", another from Kingston by way of Marbletown to the same place, and one down the East Branch of the Delaware to Papacunck", and no other roads in or leading to the Delaware County Area."

"Prior to 1785 Hans Ousterhoudt and others cut out a road and marked and numbered Mile Trees, one to thirty-five, along the same from the widow Cole's house at Lakawack to John Shaver's house at or near Pepacton. George Metcalf who went over the road in September 1785, called it "the common road, on which were sundry mile trees numbered from Lackawack to Pawpacton." The streams crossed by the road were the Beaverkill, Shin creek, the Wollowemoc, and the Neversink. James Chambers who had been over that region with parties in the Revolution, acted as guide for the party with which Metcalf traveled. He tells us that road was "formerly measured and cut out by Hanse Ousterhoudt and others", and that the mile trees were "marked from one to thirty-five beginning at the widow Cole's house in Lackawack and ending near John Shaver's house at Pawpacton." The existing road over Beech Hill is doubtless part of the same road. Probably the road was the route followed by the Indians and Tories in the attack on Lackawack in July, 1778, and by the militia in the pursuit of Ben Shanks and his Indians after the death of Colonel Graham in the ambuscade near Grahamsville in September, 1778."[3]

In the book of "Margaretville and the Surrounding Area", by Ethel Bussy Coulter, "In 1785 Hans Ousterhoudt cut a road through where the present Beech Hill road goes. It was the only North-South route across the Catskill Mountains."[4]

John D. Monroe writes in "Chapters in the History of Delaware County N.Y.", "By act of March 28, 1805, David Bostwock, Stephen Benton, Lemuel Hotchkiss and Terrence Donnelly were granted the exclusive franchise for seven years of running a stage or stages over the turnpike between Catskill and Unadilla; stages to run at least once in every week, and the fare to be five cents per mile per passenger, and the same rate per 150 pounds of excess baggage."

Taverns were first licensed in Franklin in 1797. After the year of 1812, Franklin had numerous distilleries. "Thousands of barrels of whiskey were sent from Franklin to Philadelphia on rafts; as many as one hundred barrels are said to have been carried on one raft."

"In 1840 the Middletown and Delhi Turnpike Company was incorporated to build and maintain a turnpike from the top of Pine Hill to Delhi. The road, 33 miles in length, was completed in 1849 at a cost of $485 per mile. For the most part it was constructed in the old roadbed, but at and south of Margaretville a new road was built on the west side of the river, and it was brought into Delhi by a new road on the east side of the river there, and a new bridge was built across the river at Delhi in 1848. Before this the route of the Delhi-Kingston road was by way of Sherwood's Bridge. The Middletown and Delhi Turnpike Company abandoned its franchise from Pine Hill to the north line of the village of Andes in 1895; but the residue was operated as a toll road until 1906, when, after certain litigation, the franchise was conveyed to the County of Delaware for $4,000."[5]

Henry Edgerton of Delhi ran a stage coach line from Delhi to Kingston. He had nine passenger Concord coaches with a large rack on back called a "boot" for carrying express and baggage.

"From Delhi to Roundout the time was scheduled for 16 hours, and the fare was $2.75. From Kingston to New York it was $3.50, making a total of $6.25 to ride from Delhi to New York City."

Coaches left Delhi at 2 A.M. for Roundout. They reached Roundout in time to meet the boats.

One driver, "Cage" Corbin, who was known to run his horses, had come from Kingston. At the Andes Stage Stop and Post Office, he went into the station with a sack of mail. While he was there one of his horses dropped dead. A man standing nearby saw the horse fall and went inside to tell Mr. Corbin about the horse. "Cage" very casually replied, "Yes, I know, he died on top of Palmer Hill, but he didn't have time to fall down."

Another story of "Cage" when he was driving the stage coach from Delhi to Kingston mentions that they had to pay a toll every few miles. The stage company was supposed to pay its tolls quarterly, but they were behind in payments, so below Margaretville the toll keeper wouldn't let "Cage" through. He is said to have unhooked his lead horses and with a chain, the gate was torn off and dragged away. Afterwards he hooked his team up and went on.[6]

In "Chapters in the History of Delaware County New York", John D. Monroe writes, "In 1800 the population of Delaware County was 10,228, and there were 1,681 taxable inhabitants in the county. Total assessments for taxes in 1798 were $85,043.75. The area of the county, with present boundaries, is 1,449 square miles, or 927,360 acres, the 1940 population was 40,989, and the number of farms was 3,738. The assessed valuation of the entire county in 1947, as equalized, was $46,519,599.10, showing an increase in assessments for purposes of taxation from 1798 to 1947 at the rate of $311,641 per year. The rate of increase of families from 1790 to 1940 has been at a rate of about 73 per year."

Delaware County had 2358 miles of highway in 1925. This was the third greatest county mileage in the state.[7]

As trucks and cars came into use, better roads had to be built and macadam roads replaced dirt roads along main routes.

The following information of the building of New York State Route 30 comes from a letter from Foster Beach. His father was the engineer in charge of construction of the state highway from Downsville to Shavertown in the late 20's. The field office was set up at Pepacton, and later when the highway was built from Shavertown towards Margaretville he was also engineer for New York State Department of Public Works on that project about 1930 and the field office was at Shavertown.[8]

In a letter from Robert Hoag of Livingston Manor, N.Y. to David Hoag and All, dated March 10, 1936, ..."They say that spring is just around the corner but we remember what the weather was 48 years ago this week (referring to the Blizzard of 1888) up on Perch Lake Hill (town of Andes) and if they should get a storm like that now the younger set would have something to talk about. We broke out our own roads then and thought nothing of it. Now it takes a lot of the tax payers money to keep the roads so the automobiles can run. It will cost this town $3,000.00 this winter for plowing the roads and I guess that Andes will be just as bad off."

Another letter enclosed with a Christmas card postmarked December 21, 1936, from Robert Hoag of Livingston Manor, N.Y. to Christina Hoag, "Just a few lines in here to say that mountain of ice and mud called Beech Hill separates us for several months each winter"...

Again referring to Beech Hill road in the town of Andes, Robert Hoag of Livingston Manor, N.Y. writes to George Hoag and All on January 26. 1940, "I hear that Beech Hill is going to be put on the map and a lot of the Delaware County travel will come this way to get on Route 17."[9]

THE MIDDLETOWN AND DELHI
TURNPIKE COMPANY TOLL GATE
AT LITTLE DELAWARE,DELHI

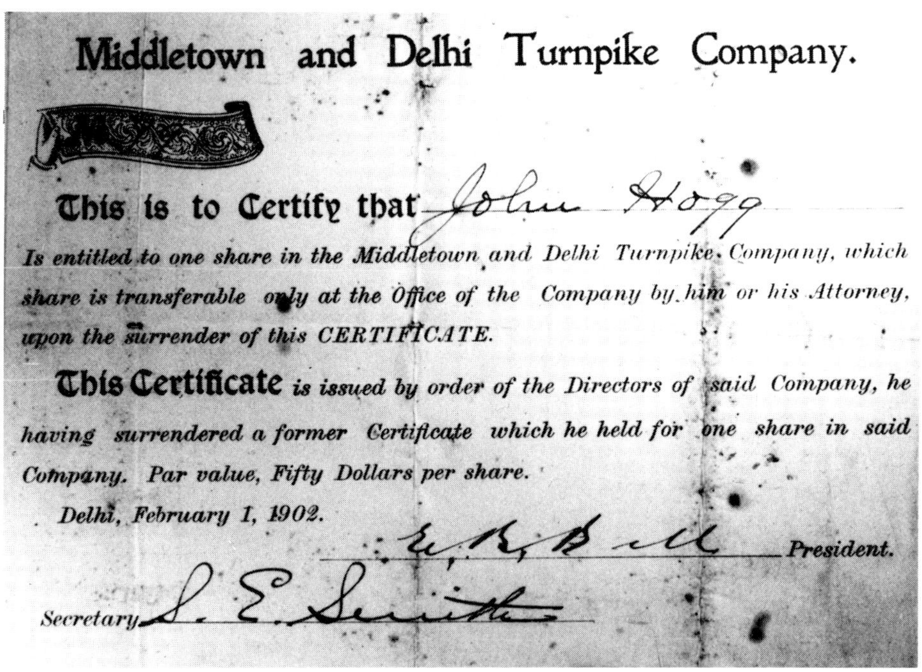

Middletown and Delhi Turnpike Company.

This is to Certify that John Hogg

Is entitled to one share in the Middletown and Delhi Turnpike Company, which share is transferable only at the Office of the Company by him or his Attorney, upon the surrender of this CERTIFICATE.

This Certificate *is issued by order of the Directors of said Company, he having surrendered a former Certificate which he held for one share in said Company. Par value, Fifty Dollars per share.*

Delhi, February 1, 1902.

President.

Secretary S. E. Smith

Little Delaware Toll Gate

The Toll Gate near the Little Delaware. Delhi was operated by John and Margaret Hogg (later their last name was spelled Hoag). John was born in Hawick, Scotland and on December 14. 1849 he married Margaret Fraser of Hobkirk, Scotland. They came to the United States "Lading ship Shannon left Liverpool June 2, 1851 arrived New York July 17. 1851" with their first child James, who was 9 months old when they left Scotland. They brought a few personal items and some carpenter tools. They first settled in the town of Andes as noted in the 1869 Atlas of Delaware County, N.Y. by F.W. Beers. page 30 in Dist. No. 21 not far from the Beaverkill. As mentioned in the Delaware County Biographical Review, page 572, he became engaged as carpenter. In later years he moved to Delhi where he tended the Toll Gate for the Middletown and Delhi Turnpike Company. They had a total of twelve children. In those days the families were often large and some of the children died young. Three of their children died - Christy at age four in 1859. Willie at age three in 1860 and one infant in 1869. On August 19. 1883 another son. Frank died at the age of 21.

There is a humorous story of the Toll Gate as one day while John was tending the Toll Gate a customer arrived and immediately disputed the toll which involved an extra ½ cent. It had been customary for the Toll Gate Keeper to keep a whole cent for the extra ½ cent since there was no ½ cent coin. John said. "Wait here". and he took the penny inside and cut it in half and returned with one half of a penny for the customer!

In later years, the Toll Gate building was moved and turned around and remodeled. It was used as a residence by some of John's and Margaret's descendants for many years. It still stands at the intersection of Federal Hill #1 and Route 28. There are also two houses still standing near that area that John had built. One of them at the intersection of the Glenburnie Road with Route 28 just next to the Glenburnie stream and the other house is located on up Route 28 towards Delhi and just beyond Federal Hill #1 on the left side of Route 28. It is presently owned by Lloyd Georgia.

The following poem written by eight of the children of John and Margaret Hogg (Hoag) tells much of the life and wealth of this family.

A Christmas Present for Mrs. Hoag

Mother, here's a Christmas gift
That we for you have bought
And it contains far precious truths
Which you to us have taught.

We your children all are here
At our old home once more
On Christmas Day in this good year
of 1884.

James is here, the boy you brought
From Scotland's sunny shore
And many a thousand miles since then
His feet have wandered o'er.

John and Robbie both are here
And Will and Davie too
We all have come our parents dear,
To spend this day with you.

Ella we meet and kindly greet
And Jennie by the way
And Tommy, youngest of us all
We all are here today.

And Frank dear boy we'll not forget
For 'twas his purpose too
If he had lived, to help and get
This Christmas gift for you.

Long may you live our Mother dear
To read and trust this word
And pray that we your children here,
May learn to love your Lord.

May there for you adorn life's path
The sweetest flowers bloom
And Jesus, be your stay and staff
Dear Mother to the tomb.

Your children
Autumn Leaves   December 25, 1884.[10]

Mrs. John Hoag died on Monday, January 22, at the home of her daughter, Mrs. Warren Dean at the toll gate three miles from Delhi, aged 75 years, 9 months and 10 days.

The funeral was held at 11 o'clock Thursday and intermentat made at the Flats cemetery.

Mrs. Hoag had misfortune to fall in September and break her leg and has been grdually failing since that time. Besides her husband she is survived by eight children—two daughters Mrs, William VanKeuren, of Andes, Mrs. Warren Dean, of Delhi, David Hoag at Shavertown, Robert Hoag, of Lewbeach, William Hoag in Glenbunie, John Hoag at Big Pond, Thomas Hoag near Lake Delaware, and James Hoag in the west.

Her maiden name was Margaret Frazier and she was born in Scotland, and there married John Hoag. A few years latter they came to America and arrived here June 17, 1850, and have since that time resided mainly in the towns of Andes and Delhi. For a number of years theykept toll gate on the Andes and Delhi turnpike.

January 22, 1906

SUGGUMBED TO WEIGHT OF YEARS

John Hoag, for Many Years Keeper of the Toll Gate, Passed Away April 17, 1814

John Hoag, in his 92 year,, died from old age Friday, April 17, at the bome of his daughter, Mrs. Warren Dean at the Toll Gate on the Little Delaware.

He was born in Scotland and came to America over sixty years ago. Of the sixty on the vessel with him only three survive, Mrs. Currie, Mrs. Foreman, and Mrs. Cooke, in Bovnia, all sisters.

His wife died eight years ago. He is survived by six sons, David, at Shavertown; Robert, at Livingston Manor; Will, at South Kortright; John in Hardenburg; Thomas at Lake Delaware, and James in Pennsylvania, and the daughter where he died.

April 17, 1914

The Delaware & Northern Railroad

The Delaware and Eastern Railroad came to the East Branch of the Delaware river valley from East Branch to Arkville in 1906. In 1911, as mentioned in "The Delaware and Northern and the Towns it Served", by Gertrude Fitch Horton, under foreclosure proceedings the railway was sold. It was reorganized and incorporated as the Delaware and Northern on October 14, 1911. As railroad business kept dwindling a motorized brill car was purchased in 1926 and used for passenger, express and post office car until 1942 when it made its last trip.

The brill car was called the "red heifer" because of its dull red color and its whistle sounded like the "blat" of a heifer.

"Right from the beginning of the road they recorded 140,000 paying passengers per year, and this was almost all local people."[11]

On April 24, 1990, Bob and I spent a most pleasant afternoon, along with Bob's Father, David Jacobson and his Aunt and Uncle, Dorothy and Isaac Jacobson, at the home of Mabel and Clara Miller, talking about the earlier days of their youth around Pepacton, N.Y. The following question from Bob brought much laughter from David and Mabel, "You remember the first train at Pepacton?" Mabel replied, "Yes, I think we do don't you David?" David and Mabel were walking down to Ed Green's (later the property was owned by Rube Johnson) for groceries. There was a train across the road when they went down and there was a man that helped them across the top of the car. "When we came back there was no one there. They was way down the road working. We stood and waited a while. Nobody came, so Dave and I crawled under the car!"

In 1919 while Ike and David Jacobson were working on the railroad a large snowstorm produced four feet of snow and they could not get back from East Branch. They chopped a hole in the ice to get water for the engine. Ike Jacobson recalled that he and Tommy Divine stayed at Harvard.

While Myers Jacobson was night watchman on the railroad in Downsville the steam engine had to be run up the track to the water tank for water. Dr. Brittain was there and wanted the privilege of running the steam engine. It was said to be a very fast trip with Dr. Brittain running the steam engine!

Dr. Brittain had a horse that kicked so he had extra long fills put on his two wheeled cart so the horse could not kick him.

Charlie Tiffany and Charlie Hawley were hurrying to a ball game and at the farm last owned by Charlie Allen at the intersection of Back River Road and Miller Hollow there was some flat cars ahead of the engine. Their view was obstructed by a board fence. They thought they could beat the train but both were killed along with their horse.

The railroad was busy from the time of the first train in 1906. It hauled mail, freight and passengers. There were two freight trains and two passenger trains running up and down the valley each day.[12] From "Damn Nuisance", Harry D. Archer writes, "During the 1920's....the same old story for all railroads, business fell into the hands of the truck and the automobile."...

"Coming over the horizon and ever getting closer was the creeping thirst of New York City. Croton and the Roundout were not enough and by 1914 the Ashokan Reservoir was built and seventeen miles of the Ulster and Delaware were relocated. 1923 came and the large tunnel of the Gilboa dam was now sending water to the Ashokan and still, they were looking for more water. And now eyes were roving over the East Branch Valley and by 1928 most everyone from Downsville to Margaretville had a pretty good idea as to how far up a hillside the water of the new dam would reach and how far up the hollows the water would go."...

One Samuel R. Rosoff, known as "Subway Sam," purchased the railroad and took charge on January 1, 1929. Mr. Rosoff managed to keep the railroad running until a settlement was made with New York City in the late summer of 1942."

"In the summer of 1920, the Vitagraph Motion Picture Company scouted the D. & N. for suitable filming of a picture based on happenings of a mining and lumbering railroad in the wilds of Alaska. The name of the picture was "Single Track" based on a book by that title. This film and all others belonging to Vitagraph were destroyed when Warner Brothers took over in the 1920's."

After scouting the area, the D. & N. was selected as the place for filming and so headquarters were established at the Pocantico Inn at Margaretville.

"Moore's trestle was a weather-beaten hemlock affair and by 1920 it was getting the eye of the State Bridge Inspector. The movie company wished the trestle more photogenic so they sprayed it with a clay and water mixture. Shortly after the movie activities were over with, the Bridge Inspector came through and was most happy to see what he thought was a new wooden trestle."[13]

This wreck as the train neared Arena trapped Clare Cowan for five hours in the East Branch of the Delaware on May 24, 1908.

From top to bottom are views of the Pepacton Railroad Depot, Shavertown Railroad Depot, Union Grove Railroad Depot and the Railroad Depot at Arena. All of this area now lies beneath the Pepacton Reservoir.

Moore's Railroad Trestle which crossed the Tremperskill Stream. The church steeple in the background at right of photo is that of the Pleasant Valley Methodist Church.

Copy of Freight Bill is from the collection of Mabel and Clara Miller.

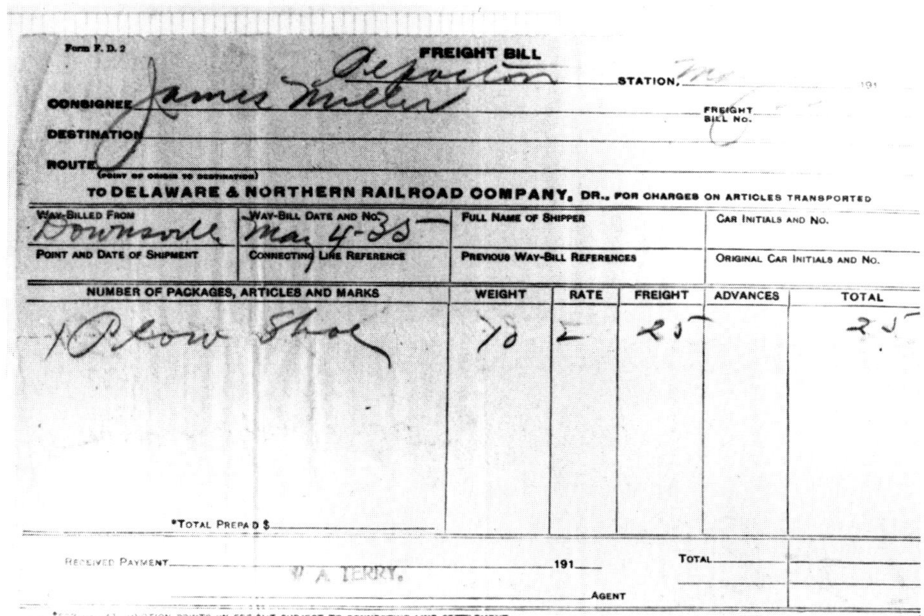

Form F. D. 2

FREIGHT BILL

Arena STATION, ... 191..

CONSIGNEE James Miller

FREIGHT BILL NO.

DESTINATION

ROUTE (POINT OF ORIGIN TO DESTINATION)

TO DELAWARE & NORTHERN RAILROAD COMPANY, DR., FOR CHARGES ON ARTICLES TRANSPORTED

WAY-BILLED FROM	WAY-BILL DATE AND NO.	FULL NAME OF SHIPPER	CAR INITIALS AND NO.
Downsville	Ma 4-35		
POINT AND DATE OF SHIPMENT	CONNECTING LINE REFERENCE	PREVIOUS WAY-BILL REFERENCES	ORIGINAL CAR INITIALS AND NO.

NUMBER OF PACKAGES, ARTICLES AND MARKS	WEIGHT	RATE	FREIGHT	ADVANCES	TOTAL
1 Plow Shoe	70	2	25		25

*TOTAL PREPAID $..........

RECEIVED PAYMENT .... 191.. TOTAL ....

W. A. TERRY, AGENT

*[illegible] AT JUNCTION POINTS ON FREIGHT SUBJECT TO CONNECTING LINE SETTLEMENT.

Post Card view of the D & E Depot at Arena.

The location of the above sawmill is uncertain. The man sitting at right has been identified as Vincenzo Presutti more commonly known as 'Jimmie Hammer'. The man sitting at left holding the peavey is identified as Vern Fuller. A peavey is used to roll logs.

Photo comes from the Delilah Babcock Collection.

S A W M I L L S

The Youngs family that operated a sawmill on the first farm up Gulf Road, Town of Andes, N.Y.
Doris Allen Collection

Portrait of Steve Brown, operator of a sawmill on Holiday Brook. Part of the stone wall dam still stands on the site of the former Steve Brown sawmill along Holiday Brook. This mill burned to the ground. He then moved to Arena where he operated another mill.

Mabel and Clara Miller Collection

Arthur Shaver Sawmill at Pepacton.

Doris Allen Collection

Sawmills

"As the county was developed by the labor of these industrious and intelligent pioneers the whole face of a neighborhood became transformed. The forest was crowded back and in its place appeared smiling fields of wheat and rye, corn and buckwheat, hay and potatoes. Apple orchards, plum trees, and currant bushes appeared on every farm. The log-house and barn gave place to frame buildings; horses displaced oxen in many of the services of the farm and the family. Roads were laid out and maintained throughout the county. Mills for grinding grain, which at first were few and distant, were erected at convenient places on streams which furnished water power. When these mills were well nigh inaccessible the pioneers had recourse to home made wooden mortars, which were dug out of a green stump large enough to hold a peck of grain. Over this was bent a tough sapling to which was tied a heavy wooden pestle. With this rough apparatus the farmer could break the husks from the grain, and even crush the kernels into a kind of rude meal.

"Saw-mills were early introduced at many suitable mill-sites. These were generally erected near pine or hemlock forests, and lumber was cut by them for the new frame houses and barns which everywhere began to be erected.

"For many years lumbering was one of the great industries of Delaware county. At many places both on the East and West branches of the Delaware river great rafting stations were maintained. The lumber was cut in the winter, and either prepared for rafting unsawed, or it was sawed into boards and joists and scantling. In the spring this lumber was built into rafts in a protected eddy of the river. Then the lumbermen taking advantage of the usual freshets of the spring started their rafts on the river. It was no easy task, and not wholly without danger, to steer the rafts through the rough and sinuous current and past the sharp headlands and rocks. When the narrow part of the stream had been passed, usually below the junction of the two branches of the Delaware, the smaller rafts were joined together, four of the former making one large raft. In this fashion the raft was run down the whole length of the river to the great lumber market of Philadelphia. It is only necessary to add here that the lumber of Delaware County has long since been exhausted and instead of the supply being sent out in rafts by the rivers, it has now become necessary to bring it in by the car load on the railroads." (This was written in the year of 1898 in "Delaware County N.Y. History of the Century 1797-Centennial Celebration June 9 & 10, 1897" Wm. Clark Publisher, Delhi, N.Y. 1898.)[14]

When Delaware County was established the present town of Andes was a part of the Town of Middletown. By act of April 13, 1819, part of the town of Middletown was taken to form the Town of Andes. The name was selected because of the mountains within its borders. Mt. Pisgah being the highest peak at 3,480 ft.

The first town meeting was in March, 1820, in the village of Tremperskill which name was later changed to Andes in 1821.

"Israel Chapman built the first gristmill a short distance below the present bridge on Cabin Hill Road. A few years later John Vaughn opened another gristmill on the site which later was used by the Dowie Mill, below the village. Mr. Vaughn also built the first sawmill in the town in 1811."

"In the days when the hemlocks grew in abundance on the side of Mt. Pisgah and the other lesser mountains around Andes, John Christian Bohlmann was one of the pioneer tanners in Delaware County.

"Born in 1812, in Neurenberg, West Prussia, Mr. Bohlmann settled in Andes shortly after he came to America. In 1839 he took over the old stone cooperage which had been built at the head of Main Street eight years previously.

"Turning this structure into a tannery Mr. Bohlmann tanned leather for manufacturers in all parts of the country until the supply of hemlock was depleted. Mr. Bohlmann died in 1889, but he is well remembered by the older residents of Andes.

"One of the beauty spots of Andes is Bohlmann Memorial Park which adjoins Bohlmann Library. Both library and park were given to the town by Mrs. Hugo Gorsch in memory of her father - John Christian Bohlmann one of the pioneer tanners of Delaware County."

H. M. Seamann had a tannery at DeLancey for many years and later as the supply of bark diminished he operated a gristmill at the site.[15]

Ethel Bussy Coulter writes, "The first settler in the valley in 1784, where the village of Margaretville now stands, was "Ingos" or "Egganaus DuMond" as he signed himself on the town records. He was a nephew of Harmanus DuMond.

"He sold his claim later for $100.00 to John Tompkins, who built the first sawmill at this point.

"Tompkins was succeeded by Salmon Scott and Jephtha Seager who divided the land. Scott took the mill and the west half of the farm and Seager took the remainder.

"Scott was succeeded in 1843 by Dr. Orson M. Allaben, and Seager sold to David C. Cluyter.

"A Mill house had been built near the mill, and Salmon Scott had built a dwelling across the flats. These two buildings and the mill were all that were standing in 1843 where Margaretville village now lies. The sawmill made way for the enterprises and industries which followed.

"The road down the river was opened from the sawmill as a log road in the winter of 1843-4."

There were thirteen sawmills on the Millbrook during the peak of lumbering activity.

The Shaver brothers erected a sawmill on the Tremperskill above Shavertown. Crude plank houses were constructed.[16]

Jonah Hotchkiss operated a sawmill near the center of what became Shavertown village. A dam was thrown across the river to supply power. His dwelling was just behind the Hulbert store and post office. The Hulbert store and post office was still owned by Amanda Hulbert (Fletcher) at the time of the construction of the Pepacton Reservoir but the store was operated by Jake and Inez Atkin and Inez was post mistress. It had been known as Atkin's store and post office for many years. The house where Jonah Hotchkiss lived was the oldest house in Shavertown at the time the Pepacton Reservoir was built and it was last owned by Woodrow Dent. The house was around 130 years old.[17]

"The first sawmill in town (town of Colchester) was built by June and Denham. It stood on the binnacle below the Mary Jane Shaver house at Pepacton, and was driven by water power from the river. It was used to cut up pine timber.

"William Horton tanned the first leather that was manufactured in Delaware County."

"John and Jacob Radeker came in 1795 from the vicinity of Wallkill, Ulster County. The brothers bought 400 acres of new land near the hamlet of Colchester. Here they built a sawmill, a gristmill and a carding mill. The carding mill was, for many years, the only one of its kind for miles around."

John More came as a single man and worked in the crude sawmill of William Early, located across the river from Cat Hollow. This mill had a carriage which had to be run by treading it with a foot."[18]

The 1869 Atlas of Delaware County New York, from actual Surveys by and under the direction of F. W. Beers assisted by A. B. Prindle and others, published by F. W. Beers, A. D. Ellis and G. G. Soule lists many sawmills. Among them is a sawmill and tannery at the site of More's Falls along the Tremperskill in the town of Andes.[19] This area is now part of New York City Board of Water Supply lands. The falls appear from time to time as the water of the Pepacton Reservoir drops to lower levels.

"Although lumbering and farming, with their related industries had always been the chief occupations in the town of Colchester, a few other ventures were undertaken. In the 1880's, among them was the raising of tobacco by David and Joe Rothensis, on the Warren and Cable flats across the river from Downsville. The drying house, located in the center of the field, had beams running lengthwise, over which the tobacco leaves could be thrown to dry. This house contained many windows which were always open for wind circulation. The dried tobacco was transported to the former Hillis Building (now owned by Charlie Leins) where it was made into cigars. This cigar industry, which started in Downsville on a small scale, later developed and was moved to Middletown, New York."[20]

The Youngs family operated a sawmill on the first farm up Gulf Road in the town of Andes. This farm was in more recent years owned and operated by Isaac and Dorothy Jacobson and later subdivided by them.

Steve Young ran the sawmill and his son, Elmer Young was the sawyer. Logs came to the mill from a tram road which ran along the mountains almost to Beech Hill. The tram road had wood rails. Cars loaded with logs were pulled by oxen. The oxen worked between the rails. They sawed ten thousand board feet of lumber per day at this sawmill. Over one million feet of lumber was sawed by steam power. Myers Jacobson stayed in a small building at the mill site and at 5:00 A.M. he was up to build the fire for the steam engine. His son, David Jacobson, also worked at this mill.

The lumber was hauled to the railroad in Pepacton. Some of the teamsters were Jim Miller, Rob Miller, Tom Brown and Bill Matthews.

There was another sawmill on the same farm located near the Andes-Colchester town line on Holiday Brook road along the Holiday Brook stream. You can still see part of the stone wall which made the dam for the water power. Later it was run by steam. This mill was operated by Steve Brown and logs came in off the tram road, and Anslee Teams, a Roscoe business, hauled the logs from the tram road to the site of Steve Brown's mill. The horses were kept at the Shaver farm.

Some of the items made at this mill included lawn swings which sold for $4.00 and clothes dryers which sold for 50¢.[21]

A newspaper clipping from the Downsville News, dated November 2, 1916 reads as follows, "The News in Arena -

"S.W. Brown is still booking orders for lawn swings, and all kinds of turned goods. He did a large business in this line before his mill at Pepacton was burned to the ground. But of late he has not been operating his mill as, in a recent accident, he sustained a broken rib and other injuries,"...[22]

Rob Miller was skidding logs for the sawmill on the Will Shaver farm when a limb caught and it was drawn around. The limb then broke and came back breaking Rob's leg. The Doctor set the leg but when it healed it was crooked leaving him crippled for the rest of his life.[23]

Post card view of Cox's Mill, Union Grove, is from Dorothy Andrews Collection.

The above sawmill was located on the Will Shaver (later John Shaver) farm on Holiday Brook. Seated from left to right are Ernest Hanes, Adam Cammer, Geo. Redmond, the mill owner ___ Morterstock, ___ Sprague, Andy Miller, __ ___ , Myers Jacobson is sitting on the stone wall at right.

Mabel and Clara Miller Collection.

George Robert "Rob" Miller is seated on the wagon and the horses from left to right were Roy and Rob.

Mabel and Clara Miller Collection.

The site of the Bill Murphy sawmill was located just beyond the present day Beaver Pond in the above photo.

The Bill Murphy sawmill was located about one mile from the Holiday Brook road beyond an area called the swamp on the Myers Jacobson farm.

There was 600,000 board feet of lumber sawed at this mill. The lumber was hauled by the Jacobson family from the sawmill to Pepacton and then it was loaded on the railroad cars.

One area near the sawmill that was to be cut for logs was too steep for horses. David Jacobson and Caleb Bush built a log chute down this steep area. The chute was made of two small logs in the middle and two large logs on the outside. The logs were fastened by stakes and spikes. The logs were placed in the chute and they would go down this steep area almost to the sawmill. Oscar Scoville lived near this mill site at one time.

The area now is part of the New York State Forest Preserve.

These two photos and the information below come from newspaper clippings in the scrap book collection of David Jacobson.

"Floyd Woodin of Andes submits these photos of the former Howard and McNair sawmill located on the Tremperskill Road where the Oscar Babst estate (former Walter Liddle property) is located. Mr. Woodin says that he had never seen an operation such as this where a slide a mile and a half long, came down the hillside. The slide was hewn of logs and V-shaped. A man named Cy Far walked along it and greased it each day to give it good skidding properties.....Logs were shipped by the D & N Railroad."

Bluestone Quarries

"Early settlers at the eastern base of the Catskills found a hard, blue-gray stone breaking through the earth and forming long parallel ledges which often alternated with bands of soft, red shale. Later on, geologists traced these ledges to Upper Devonian days some two hundred million years ago. They found that at higher elevations among the Catskills the bluish sandstone turned gray, green, or pinkish and lost something of their hardness and strength", writes Alf Evers in "The Catskills from Wilderness to Woodstock". Bluestone was used by the early settlers as grave markers. Later on marble became more fashionable for grave markers. Bluestone was used for steps or stones at front and back doors and circular pieces were cut to make covers for crocks. Pickles and saurkaraut were kept in these crocks. Bluestone was also used for fireplaces, chimneys and cellar walls.

In the 1840's as economic growth changed for the better, bluestone was sent by wagonloads to New York and other cities for flagging for sidewalks. The stone was put on sloops and sent down the Hudson River.

"By 1850 sloops laden with bluestone flagging were leaving the wharves of Hudson Valley trading towns not only for New York but also for Boston, Philadelphia, Baltimore, Charlestown and other cities up and down the Atlantic coast. Before the end of the century, sidewalks of North River bluestone were echoing under the feet of people in San Francisco, Milwaukee, St. Louis, and Havana, Cuba."

Margaretville, Roxbury, and many points between had thriving quarries and trackside stone yards, along the Ulster and Delaware Railroad.[24]

In "Pioneers of the Beaverkill Valley", Lena O. B. Tiffany writes,

"In New York City at City Hall there was placed a stone taken from a quarry at Pond Eddy. It was placed on a raft and taken down to Philadelphia, then on a boat through a canal to New York City.[25]

When the Pepacton Reservoir was built, Lloyd Terry of Shavertown moved to the Tremperskill. He built a stone building where he operated a bluestone business. Many loads of bluestone were brought there by area stone cutters and then it was sent out by tractor trailer loads to far away places. There have been many small stone quarries in the area.

The Johnston and Rhodes stone quarry in the photo above was on the Kezia Williams place near Downsville. Wallace Clark is standing in the background at right.

Both photos come from the Wallace Wynkoop Collection.

The photo below shows the stone loading dock on the Kezia Williams place. The D & N Railroad track is at left of the photo. The Kezia Williams house is shown at the top of the hill.

Hide-Out of Draft Dodgers

From the notes of my father, George F. Hoag, "Many years ago a cave near the ledges which were a part of the Harry Lakin farm at the time of the building of the Pepacton Reservoir provided a place for three local men to hide for days to escape the army during the Civil War. It also provided a hide-out for a counterfeit before that time."

A nephew of one of the three soldiers had shown the cave to George Hoag around the year of 1918. It was a small cave, the front had been walled up. Once when I was a young child, I can remember walking up to the site of the cave and ledges with my family. In my father's notes, "Later the Union Officers captured one of the men in the night at his home. The other two fled to Canada." [26]

From the ledges one could look down the East Branch of the Delaware River over the intersection of the Beech Hill Brook with the East Branch of the Delaware River and beyond towards Pepacton. At the time of the building of the Pepacton Reservoir it over looked as one looked down river first the George Hoag farm, Ina Seath property, a farm operated by Harold Paul and the farm owned by Don Bramley (These farms had large fertile flats along the river). The Ina Seath property included the old store building that had been Shavertown's first store operated by Richard Woolsey and later it had been Andrew Finkle's store, not in operation in later years. From these ledges one could also look up river over property last owned by Forest Lakin, the Catholic Church, which had first been a Presbyterian Church built in 1851, and the Shavertown Cemetery across the road from the aforesaid Church. In the History of Delaware County, N.Y. by W.W. Munsell 1797-1880, page 111, "The Cemetery near the church is on the site of the old Indian burial ground. The bones and hunting knives of the savages are frequently exhumed, and in several cases wrought nails have been found, with which it appeared some rude boxes had been made for the more illustrious of the red men." Beyond the church and cemetery one could see the Harry Lakin farm, the Byron Huggins farm, the Don McCune farm across the river and the buildings along Shavertown's River Street.

This seems like a likely place for a hide-out since there is a most magnificent view of the valley and the river from this area.

In one place on the ledges appears the initials "G.F.H. 1923" carved by George F. Hoag in the year of 1923. Just to the right of these initials more faintly carved is "D.C. Hoag" carved by his father David C. Hoag. There are also many other initials scattered among this area. Today the area provides a breath taking view of the Pepacton Reservoir.

In the 1869 Atlas of Delaware County, N.Y. by F.W. Beers, Town of Andes Map, page 30, it is noted "Cave Discovered by House 1860." This might possibly be the hide-out location as the small cave that I remember would have been very close to that area. The area is now owned by High View Gun Club.

My Aunt Edna Hulbert Goyette writes in a letter of January 20, 1990, "Yes, I remember the cave at Harry Lakin's place. Yes, it was a nice view from there I remember it. I can't tell you about the stories. I don't remember them. It's faint in my mind about the counterfeit."[27]

Copy from page 30 of 1869 Atlas of Delaware County,N.Y. by F.W. Beers, shows "Cave discovered by House 1860".

The area of the Cave discovered by House was part of the lands belonging to Harry Lakin when the Pepacton Reservoir was built. It now belongs to High View Gun Club.

The ledges on the former Harry Lakin farm now owned by High View Gun Club where the initials of George F. Hoag "G.F.H. 1923" still remain. Just to the right of these initials there is a faint impression "D C Hoag" for his father David C. Hoag. There are also many other sets of initials scattered around the area.

The photo below bears a date of Jan. 12, 1924 and is labeled "Entrance to Civil War and Counterfeiter's Cave." The dog at the entrance belonged to Letha Hulbert Hoag.

Betty Berke Collection.

The view from the ledges on the former Harry Lakin property as one looked up the river over the Forrest Lakin property, the Catholic Church and Shavertown Cemetery, Harry Lakin land, Byron Huggins land, the Don McCune farm lands and Shavertown's River Street.

This view from the ledges on the former Harry Lakin farm overlooks the George Hoag farm, Ina Seath property, the farm operated by Harold Paul and the Don Bramley farm land as one looks down the river from the Shavertown area.

The view from the ledges on the former Harry Lakin land presently gives a breath taking view of the Pepacton Reservoir.

Photo at left shows Harry Lakin's house on Back River Road below Shavertown. Dorothy Rice Collection.

The Lakin land was the setting for another incident. In 1930, Albert Lakin, Carson VanKleeck and Grant March were sitting on the porch of Harry Lakin's house. Walter Russell and Ed Russell were up in the woods in a rocky area below the ledges. They were turning up stones and somehow, Ed got a stone on his leg. Albert, Carson and Grant heard him call for help so they went up to see what had happened. They found Ed with a broken leg. They made a makeshift stretcher which they fastened together with their belts and carried him off the hill to Harry Lakin's porch. Dr. Wakeman was summoned from his home and office in Andes. Ed was given some anesthesia and the leg was set in place by Dr. Wakeman on Harry Lakin's porch! (This was in the days before X-rays and hospital care.)

In photo at right are from left to right; Albert Lakin, Walter Russell, Ed Russell on the make-shift stretcher, Grant March and Carson VanKleeck.

Grant March Collection.

School Houses

Ethel Bussy Coulter writes, "Bussy Hollow is in Pleasant Valley which lies between Andes and the former Shavertown. The Bussys that first settled there, and for whom the place was named, were ancestors of the Bussys of Margaretville, formerly Arena, Downsville and Livingston Manor. The first of the clan to start for America was John Claudius LeDroit Bussy. He was of French nobility and extremely wealthy. On the way, he stopped at one of the West Indies where a son, Lewis, was born to his wife. The latter died during the child's birth, so a nurse was procured for the infant. Later, father and son came to South Carolina and a bill of sale recorded there on February 8, 1780, shows that LeDroit Bussy sold a Negro girl named Rose to one John Hart for the sum of 12,000 pounds. That would be the purchasing equivalent of something like $150,000 today. Such transactions brought fantastic sums from wealthy men. Even George Washington was not above such dealings.

"LeDroit soon made a name for himself in America. He served as a major in the Revolution and was also chancellor in New York City for some time. Later, Lewis, the son, rode horseback to Trenton, New Jersey, and in 1816 found his way to Bussy Hollow which was named after him. He was what is now called Superintendent of Schools in this area.

"The old Bussy homestead was built in 1836; it had a fireplace mantel in the house that was carved by Lewis himself with a jackknife."[28]

When we lived on the farm one mile below Shavertown, there was a building next to the covered bridge crossing the Beech Hill Stream which my father used for a storage building. It had at one time been used for a school house, for District #6, town of Andes. My father attended school there when he was a small boy for a year or two. There is not much information available about this school but Florence Russell Huggins wrote that her mother, Essie Shaver Russell and her sister Dora Shaver Bramley used to row a boat across the East Branch of the Delaware River and when the river was high, their father, Ezra Shaver, watched them. It was a short cut from their home to the little school by the Beech Hill Road. She notes, they didn't know how to swim but Essie was good at rowing the boat![29]

Toll Gate District

School District No. 9

Little Delaware 1873-74

Belle S. Patterson 16, Robert A. Patterson 13, Thomas D. Patterson 11, Wattie Patterson 7, Agnes N. Patterson 9, Alice Shaw 19, Jennie Shaw 16, Warren Dean 16, Bell J. Mable 14, Agnes S. Mable 11, James D. Mable 9, Robert A. Mable 16, Julia Maxwell 16, Fannie Middlemist 17, Bell J. Middlemist 16, Robert McGregor 16, Agnes B. McGregor 14, Charles C. McGregor 12, Jennie C. McGregor 10, Walter R. Mable 16, George S. Mable 14.

Jennie F. Mable 11, Willie D. Mable 8, Isabell J. Irvine 9, John A. Irvine 6, Alice Patterson 8, Robert Hogg 13, Frank Hogg 12, Jennie Hogg 10, David Hogg 9, Willie Hogg 7, Maffie A. Thomson 13, Alfred Fitch 12, Charlie Flower 10, Nelson Siring 17, Robert Leal 11, Frank Leal 11, Johnnie Leal 8, Charlotte Leal 7, Minnie Leal 10, Fannie McMullen 6, Tommie Hogg 5.

Alex Mable was school trustee and W. M. Honeywell was the teacher.[30]

TOWN OF ANDES, DISTRICT No. 20.
Fall and Winter Term, 1887-8.

No.	NAMES OF PUPILS.	Age.
1	Lillie Hull	15
2	Bessie Clement	13
3	William Clement	16
4	Lizzie Liddle	13
5	Frank Liddle	10
6	Chrissie Liddle	6
7	Lizzie Granlees	13
8	Mamie Fenton	12
9	Andrew Fenton	17
10	Tina C. Fenton	19
11	Rozella Bussy	9
12	Arthur Bussy	7
13	John D. Monroe	9
14	Mary Monroe	6
15	Mamie Liddle	7
16	Elliott A. Liddle	5
17	Everett Miner	7
18	Charles Warren	9
19	Olivia Warren	6
20	William Brace	9
21	Marshal Brace	14
22	Chrissie Smith	17
23	George A. Smith	6
24	Janie Lunn	15
25	James Dowie	14
26	Mary A. Dowie	17
27	Hattie Close	17
28	Edwin K. Davis	16
29	Frank Burtch	13
30	Alexander Burtch	10

LAFAYETTE BUSSY, Trustee.

School List of Shavertown School house, Dist. No. 19.

NAMES OF PUPILS

******

Attending School in Dist. No. 19, Town of Andes, During School Year, 1901-1902.

NAME.	AGE.	NAME.	AGE.
Mary Fletcher,	13	Edna Hulbert,	7
Blanche Fletcher,	9	Jennie Hulbert,	8
Lena Miner,	13	Edith Russell,	7
Tracy Neish,	14	Mina Gladstone,	8
Jakey Atkins,	14	Maud Atkins,	6
Bernard Atkins,	12	Ida Gladstone,	5
Earl Atkins,	10	Howard Fletcher,	16
Harry Terry,	13	Daniel Scott,	15
Ida Terry,	13	Marcus Fenton,	8
Howard Miner,	12	Florence Norris,	12
Mabel Hulbert,	9	Carmie Norris,	6

Lauren Hulbert, Teacher.
Chas. B. Neish, Trustee.

The Pleasant Valley School list for Town of Andes, Dist. No. 20 is shown above. The School House in more recent years became a store and is presently known as the Tremperskill Country Store along Delaware County Route No. 1.

At lower right is a 1931 photo of the Holiday Brook School House, Dist. No. 30, Town of Colchester, from the collection of Mabel and Clara Miller.

In front, from left to right are; Angelica Jacobson, Mina Bouw Bogart, Elsie Shaver Pardee. In back, left to right are; Forrest Davis Warren, John Bouw, Albert Bouw and Billy Smith.

Holiday Brook School Picture

First row, left to right, Nellie Hotchkiss Farber, teacher, Mable Finkle, Leah Jacobson, Vernon Shaver, Edna Tiffany,Grace Shaver, Forrest Lockwood, sister Lockwood. Second row, Eliza Tiffany, Mary Tiffany, Jennie Jacobson Jackson, Helen Kelly Rode, Maggie Shaver Johnson, Libbie Shaver Robinson. Last Row, David Jacobson, Arthur Earls, John Shaver, Adelbert Lockwood, Jasper Decker.
Mabel Jacobson Miller and Anna Jacobson Taylor are absent.

Mabel and Clara Miller
Collection.

Holiday Brook School

by Bob Jacobosn

The first Holiday Brook School meeting was held on August 4, 1903 at the Will Shaver house. At this meeting officers were elected - Charlie Finkle, Chairman, Adam Shaver, Trustee, Steve W. Brown, Clerk and O.H. Scoville, Tax Collector.

This school building still stands along the Holiday Brook Road, town of Colchester.

Adam Shaver was given full power to build a school house for School District 30 on Holiday Brook Road, to the best of his judgment.

The first teacher was Nettie Warren. Some of the other teachers were Mabel Allen, Mary E. Tiffany, Libby Shaver, Harrison Bogart, Josephine Lockwood, Arthur Russell, Delbert More, _____ Moshier, Harry Wright, Goldie Conklin, Katie Brock, Nellie Hotchkiss Farber, Ethel Bryden, Ethel Hunt, Vida Williams, Emma Holmes Lewis, Edna Tiffany and the last teacher was Margaret Salton.

Some of the officers that served through the school years were: Charlie Finkle, Myers Jacobson, Adam Shaver, Steve W. Brown, Will Shaver, Geo. R. Miller, Marshall Tiffany, P.O. White, James Miller, John Shaver, Marshall Tiffany, Jr., David Jacobson, Margaret Jacobson and Isaac Jacobson.

The last meeting of the Holiday Brook School was held on May 3, 1938. At this meeting ever member voted to keep their small school but as we all know the schools were centralized and Holiday Brook students went to Downsville.[31]

School District No. 30
Holiday Brook, N. Y.

Catherine Brock, Teacher.

J. W. Shaver, Trustee.

Pupils

John Shaver 16
Vernon B. Shaver 10
Arthur Oralls 13
Leah M. Jacobson 9
Leona E. Miller 10
Mary E. Tiffany 16
Edna M. Tiffany 10
Mabel J. Finkle 11
Sarah Paul 7
Elizabeth E. Shaver 13
Grace B. Shaver 7
David J. Jacobson 12
Mabel A. Jacobson 14
Robert S. Tiffany 17
Eliza B. Tiffany 14
Ella E. Fuller 9
Jasper J. Decker 11
Matilda Paul 5

Mabel and Clara Miller Collection.

NAMES OF PUPILS

ATTENDING SCHOOL IN DIST. NO. 30, HOLLIDAY'S BROOK, TOWN OF COLCHESTER, SCHOOL YEAR 1912-1913.

Mabel J. Finkle,	-	14
Edna M. Tiffany,	-	13
Leona E. Miller,	-	13
Ella E. Fuller,	-	12
Leah H. M. Jacobson,		13
Grace B. Shaver,	-	10
Eunice M. Jacobson,	-	7
Mabel L. Banuat,	-	5
Laura E. Bound,	-	7
Vernon B. Shaver,	-	13
Willard M. Bound,	-	13
Walter J. Bound,	-	11
David J. Jacobson,	-	15
Isaac W. Jacobson,	-	9
Georgia M. Shaver,	-	6

HARRY L. T. WRIGHT, Teacher.
James William Shaver, Trustee.

Both photos are from David Jacobson Collection.

Photo above-Holiday Brook School House - 1917. From left to right, in front, Cecil Young, Nelson Hatch, Margaret Alton (10 yrs.) (later Jacobson). Second row, George Corgill, Emma Young, Georgia Shaver, Sonny Young. Back Row, Helen Corgill and Elmer Young.

The Teacher was Will Shaver's daughter, Libby.

The Youngs lived on the place later owned by Isaac Jacobson. Hatchs lived at the old Steve Brown place across the road from the farm later owned by David Jacobson and Corgills lived on the farm later owned by David Jacobson.

Holiday Brook School - 1923 - from the Collection of
Mabel and Clara Miller.

In front, left to right, Clara Miller, Woodrow Tiffany, Mary Jacobson Allen. In back, left to right, Florence Jacobson Northrup, Gracia Warren Havens, Grace Deighton, Henry Deighton and Edward Jacobson.

Holiday Brook School - 1931 - from the Collection of
Mabel and Clara Miller.

In front, left to right, Harold Youmans, Angelica Jacobson (Corgill) (Finch), Billy Smith, and Mina Bouw (Bogart).
In back, Bertha Youmans (Foreman), Clara Miller, Forest Davis Warren, Edwin Fuller, John Bouw and Albert Bouw.

Photo copy from Ann Liddle and Information from Aasta Liddle-

Brock's Bridge School about 1903

This school house was situated two miles north of Downsville. Mina Wilson was the teacher.

The girl with the white ribbon down front is identified as Nellie Reynolds, Elsie Gregory (Liddle) is the smallest girl sitting down and Jim Reynolds is the boy sitting down. He is a brother to Nellie.

Pepacton School

From the Collection of Mabel and Clara Miller.

In front, left to right, Guss Flynn, Henry Cammer and Adam Cammer. Back Row, left to right, Bertha Kinch, Mabel Allen, Nettie Kinch, Flossie Craft, Aggie Tiffany, Will Matthews, Maud Miller, Vianna Miller, Jim Miller, and Delacey Fairbairn, teacher.

Photo below is another view of the Pepacton School from the Collection of Mabel and Clara Miller.

Pepacton School - Sept. 1917
Florence Northrup Collection.

First row, Robert Misner, Edward Jacobson, William Taylor, Harley Misner, Ruth Allen, Gracia Warren, Basil Shaver, Nelson Hatch, John Hunchburger; second row, Marion Hyzer, Ruth Misner, Matilda Paul, Bernice Van Dusen, Gertrude Misner, Mildred Shaver, unidentified, Helen Hunchburger, unknown, Theresa Shaver, Daniel Sickler; third row, Macelia Fuller, Florence Warren, Sera Paul, Helen Shaver, Gertrude Beardsley, Eunice Jacobson, Harold Hyzer, Eldred Shaver, Charles Warren, Isaac Jacobson, Gilbert Warren, Wesley Allen and teacher Mary Cammer.

Vera Washburn Craft writes, "I have so many memories of the beautiful farm I lived on. In my memory, even though I was only nine when I left there, I can still picture every room in the house. I can remember every turn in the road from Shavertown to Pepacton, as I walked that road many times. I went to Church in Shavertown and to school in Pepacton."

Vera's first teacher was Edna Tiffany. Some of her classmates were Clifford Terry, Elsie Smith and the Misner children.

"Those were happy days up until the time my mother died. The neighbors were all so kind to us when mother died. She was buried Dec. 24, 1926."

The Washburn farm was situated between the Signor farm and the Floyd Davis farm. It had formerly been the Pulaski Conklin farm.

After her mother died they lived in Arena and she went to school in that two room school until she went to the Margaretville High School.

Anyone who has ever been to Pepacton, N.Y. will probably remember the large elm tree that stood by the covered bridge crossing the East Branch of the Delaware River. There is a story that Emmett Bryden once ran his car into the elm tree and he said, "Mighty! They will have to move that tree!"

The photo of the elm tree and covered bridge come from the collection of Mabel and Clara Miller.

PEPACTON

There are many stories that are connected with Emmett Bryden when he lived in Pepacton. Clifford Sprague remembers one of a traveler who stopped at Pepacton and asked Emmett the way to Downsville. Emmett replied, "Go through the covered bridge, turn left, first big town, six miles down. Everything is lovely when the goose hands high!"

The N.Y.S. Route 30 sign above was at the entrance of the covered bridge. The covered bridge connected Pepacton with Route 30. It indicates that Shavertown was 5 miles from Pepacton.

The sign is part of Chip Sidorowicz's collection.

The above photo comes from the collection of Mabel and Clara Miller.

A Pepacton correspondent writes in the Walton Reporter, on January 26, 1940, "The old covered bridge at Pepacton, which has withstood many floods and ice jams, was built in the year 1858, accroding to records in an old ledger account book of T. and J. A. Shaver, who at that time had a general store on the site where C. R. Johnson's garage now stands."[32]

Another newspaper clipping from Clara Miller's collection tells of another view, "This is the doomed Pepacton bridge, only one remaining in the United States and probably the world with a Haupt truss with arch construction. It was constructed in 1870 by the late Robert Murray of East Branch, grandfather of Mrs. Marjorie Kinch, Mrs. Harry Shepard, Miss Alice Murray, Roger Murray and Kenneth Murray of Walton. The span is 168 feet long."[33]

In the winter the bridge was a very hard pull for horses hauling loads of wood, lumber, etc. through the bridge on sleighs. Sometimes they got stuck in the bridge and another team of horses was hooked on and in later years sometimes a truck was hooked on to help pull the load through the bridge.

Jim Miller drove a 55 passenger Downsville Central School bus with students across the bridge a good many times.[34]

Ethel Bussy Coulter writes in "History and Stories of Margaretville and the Surrounding Area", Pawpachton later written Pepacton, had been an Indian settlement, the name meaning calamus or sweet flag place.

The Indians called the East Branch of the Delaware River "Papakunkill".[35]

The famous naturalist John Burroughs, who was born in Roxbury once traveled down the East Branch of the Delaware as mentioned in his book "Pepacton". He mentions Pepacton as being the Indian name for the river meaning "marriage of the waters".[36]

The post card view of Goodsir's Store above is from the collection of Mabel and Clara Miller.
Goodsir's Store was later owned and operated by Willis and Maude Terry. They owned the building until it was destroyed for the Pepacton Reservoir.

Flood at Pepacton-left to right are Jim Miller, Will Flynn, Archie Allen next to the tree, unknown, and Harley Misner.

Mabel and Clara Miller Collection.

Pepacton Ball Team

in the above photo from left to right are seated in front, Bill Hoyt, Jim Rutherford, and Chas. Hawley.
Standing from left to right, Wilbur Huntley, Geo. Kelly, ___ ____ , C. R. Johnson, George Luce, ___ _____.

Both photos are from the collection of Mabel and Clara Miller.

Pepacton Ball Team

Left to Right-in front,John Shaver, Geo. Kelly, Wilbur Huntley (Nancy Shaver's brother), in back-Geo. Luce, Claude Tiffany, Floyd Ellott, Bill Hoyt, Mary Hitt, Rubin Johnson.

Mabel & Clara Miller Collection

Post card view of Pepacton Hotel.

The Pepacton Hotel was run by the Williams'.
Some boys saw a man go in the rest room out back of the hotel so they got two boards and some wire and fastened the door shut so the man stayed in there all night!

The hotel was built on stilts to survive the floods and sometimes during the ice jams the ice would come up on the porch.

At left is a photo of part of a calendar from the business of Daniel A. Bryden.

Chip Sidorowicz Collection.

Ladies Aid Group from the Pepacton Methodist Church
Doris Allen Collection

Children in front Ella Shaver Campbell, unknown, Shirley Allen
Front Row, Lizzie Scudder Shaver, Lucy Bush VanBramer, Helen Bouw, Dorothy Townsend, Leona Miller Allen, Hazel Flynn, Second Row, Mrs. Townsend, Clara Tiffany Terry, Maggie Shaver Johnson, Mary Flynn, Agnes Allen, Clara Miller, Elsie Shaver Pardee, Mina Bouw Bogart.

Mabel & Clara Miller Collection

Leona Allen and Doris Allen at right. The Allen house at Pepacton, N.Y. in background.

In early years the Pepacton Post Office was not always in the village. Mabel Jacobson Miller first remembers the post office at the Jim Dimmick farm house. This was the second farm below Pepacton on Route 30. This farm was later owned by George Sickler.

Next the post office was at the third farm below Pepacton on Route 30. This was the Ed Green farm and the post office and a general store was located in a garage at this location.

After the railroad was built (1906) the George Holmes Milling Company built a feed store in Pepacton which was operated by Ed Green. Jim Miller worked for Ed Green at this feed store and at that time the mail came to Pepacton on the railroad and Jim would take the mail on his bicycle to the Ed Green farm where the post office was located in his garage. This was one mile below Pepacton.

Other locations of the post office in Pepacton were the railroad depot, feed store, Edison Smith garage, Emmett Bryden's place and Willis Terry's store.

It was last operated by Bob and Anne Misner in their house which was located across from the Herb Shaver farm until the construction of the Pepacton Reservoir. The Misner house was purchased by Commodore and Joyce Jacobson when the Pepacton Reservoir was built and they had it moved to a new site up Miller Hollow.

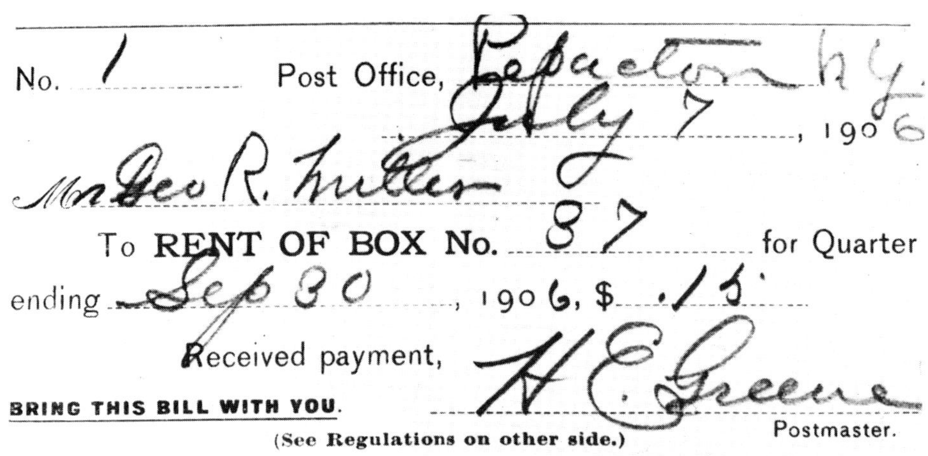

No. 1 Post Office, Pepacton N.Y.
July 7, 1906
Mr Geo R. Miller
To RENT OF BOX No. 87 for Quarter
ending Sep 30, 1906, $ .15
Received payment,
BRING THIS BILL WITH YOU. H. E. Greene
Postmaster.
(See Regulations on other side.)

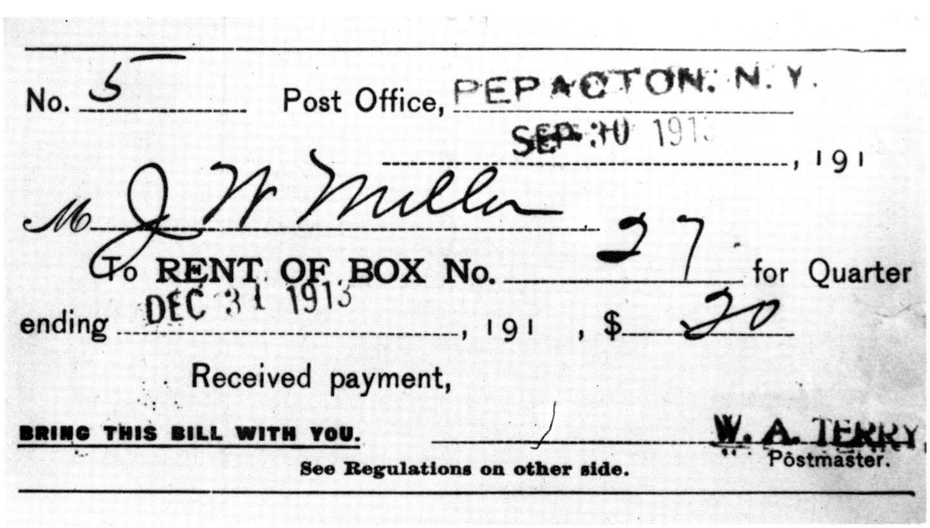

No. 5 Post Office, PEPACTON, N.Y.
SEP 30 1913, 191
M J W Miller
To RENT OF BOX No. 27 for Quarter
ending DEC 31 1913, 191, $ .30
Received payment,
BRING THIS BILL WITH YOU. W. A. TERRY
Postmaster.
See Regulations on other side.

Post Office box rent receipts are copied from the collection of Mabel and Clara Miller.

Mabel Jacobson Miller was sixteen years old in the photo at left. The year was 1911 and she went to work at Conklin's up Coles Clove. The dress was made by her mother, Ida Budine Jacobson. The material came from the Conklin Store.

Photo is copied from the collection of Mabel and Clara Miller.

Mabel Jacobson Miller has many memories of walking up the road to the Holiday Brook School when she was nine years old and her family lived one mile from the school on Holiday Brook road. She walked down the road from their home for a distance of three miles to the Ed Green place which was one mile below Pepacton for groceries and mail for the family. This means that she walked a distance of two miles round trip up the road to and from school and then another six miles round trip down the road to and from Ed Green's place giving her a total of a eight mile walk.

The store and post office were located in the garage building near Ed Green's house at that time.

In later years at the age of sixteen she went to work at Ed Conklin's. They had a general store. It was the second place on Coles Clove road from Route 30. She earned $3.00 per week. Occasionally when she had a weekend off, she would walk back home a distance of three miles.

Jennie Jacobson Jackson worked at the Scudder place at the intersection of Route 30 with the Coles Clove road. The Scudders had summer boarders. The place was last owned by Jack Bouw until the time of construction of the Pepacton Reservoir.

The building in the photo above was known as the "Bee hive". It was situated next to the store built by J.E. Goodsir and last owned and operated by Willis and Maude Terry until the time of the construction of the Pepacton Reservoir.

Photo comes from the collection of Edna Shaver.

Mabel Jacobson Miller's Memories of Pepacton

George Miller had a grocery store in the "Bee hive" building shown in the photo above before he went into the business of white washing barns. He was in the store about 1909 or 1910. Tom Hitt had the store in 1911. Hose Barnhart took over the business in 1913. He sold groceries and material and other dress making items. He built a garage in Pepacton and gave up the store about 1915 or 1916. Orley More and Anna ran the store after Hose Barnhart. There was an apartment upstairs over the store. After Orley More gave up the store, the area was used for meetings and dances. Ken Bryden played the violin, Archie Bryden played the guitar and Will Hadley played the violin for the dances.

Hose Barnhart sold Ford cars in the aforesaid garage that he built. Jim Miller bought his first car there in 1916. David Jacobson and his father, Myers Jacobson, bought their first cars there also. Hose later moved to Walton and Edison Smith took the garage over and sold Ford cars. He also made repairs on the cars and sold gasoline. Jim Miller worked for him and helped sell cars and gasoline and did repair work. Edison later moved to Unadilla and worked at the Scintilla Plant in Sidney. Gladys Bryden had the post office in the garage office after Edison moved.

Nelson Woodard had a Black Smith Shop. Billy O'Brien ran the shop after Nelson Woodard and he built a house and barn where Archie and Leona Allen lived. Hank and Walt Newman had the blacksmith shop after Billy O'Brien and he lived upstairs over the shop.

Charles Dutcher had a grocery store across the road from the garage. There was a barn in back of the store.

Jake Goodsir built a house between the Dutcher Store and his store which was known as Goodsir's Store. His store was later owned and operated by Willis and Maude Terry. They owned the building until the Pepacton Reservoir was built.

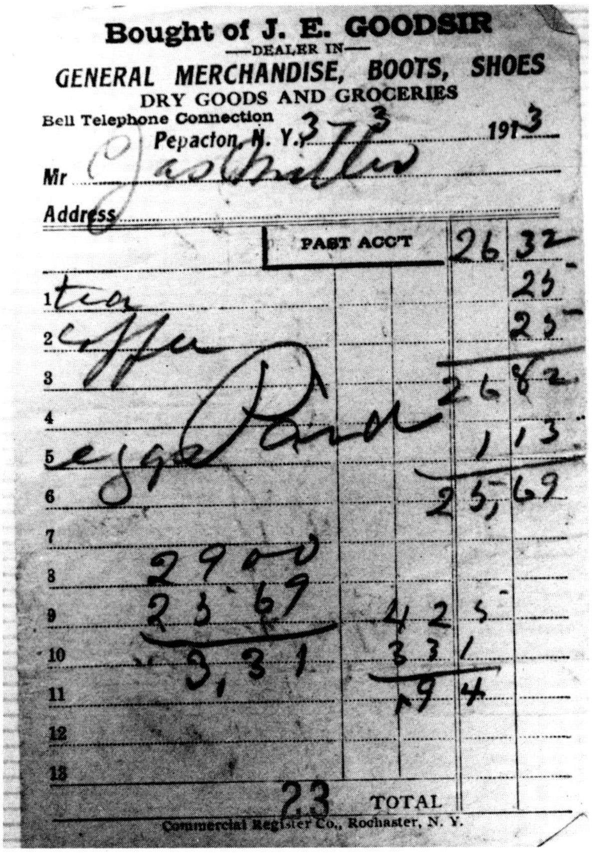

Bought of J. E. GOODSIR
—DEALER IN—
GENERAL MERCHANDISE, BOOTS, SHOES
DRY GOODS AND GROCERIES
Bell Telephone Connection
Pepacton, N. Y. 3/7/3 1913
Mr Jas Miller
Address

	PAST ACC'T	26 32
1 tea		25
2 coffee		25
3		26 82
4		
5 eggs		1 13
6		25 69

Paid

23 TOTAL

Commercial Register Co., Rochester, N. Y.

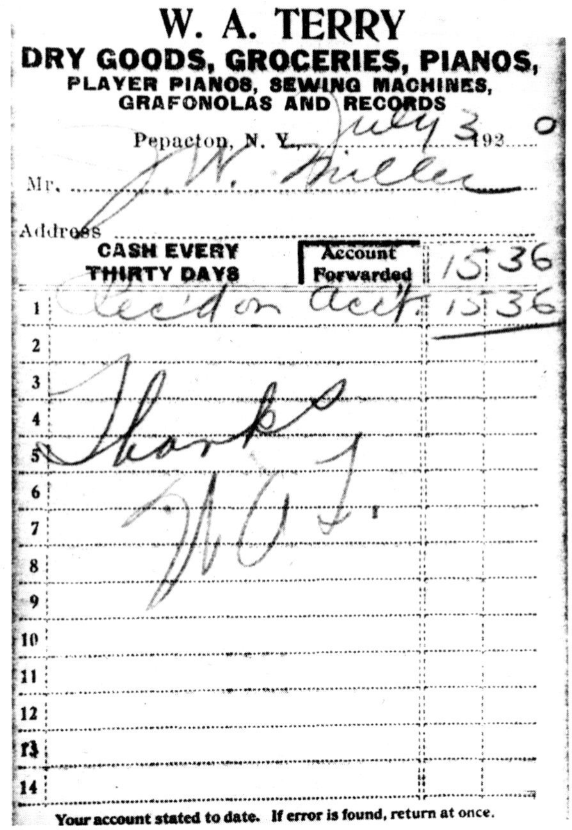

W. A. TERRY
DRY GOODS, GROCERIES, PIANOS,
PLAYER PIANOS, SEWING MACHINES,
GRAFONOLAS AND RECORDS
Pepacton, N. Y. July 3 1920
Mr. J. W. Miller
Address

CASH EVERY THIRTY DAYS	Account Forwarded	15 36
1 Rec'd on Acct.		15 36

Thanks
W. A. T.

Your account stated to date. If error is found, return at once.

These copies come from
Mabel and Clara Miller's
Collection.

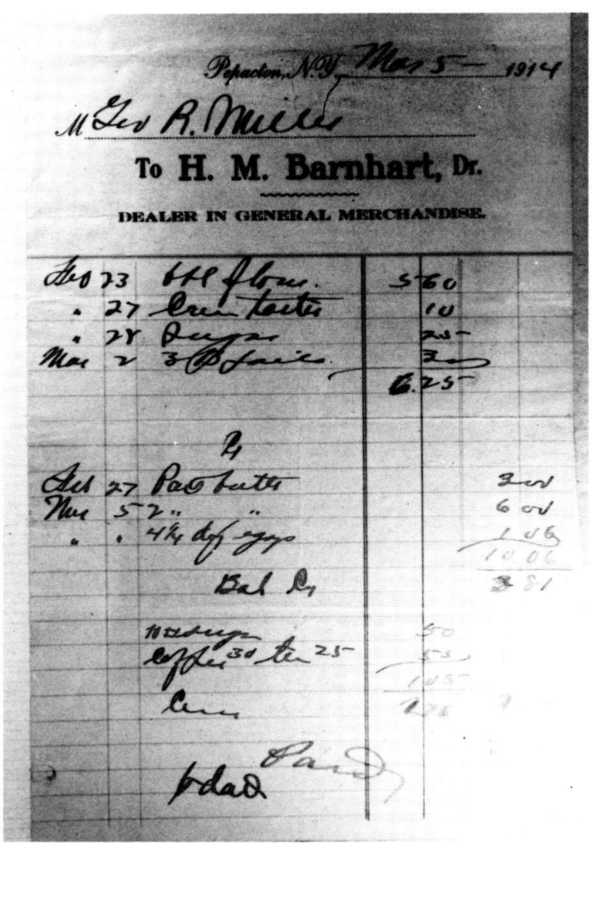

Pepacton, N.Y. Mar 5 1914
M Geo R. Miller
To H. M. Barnhart, Dr.
DEALER IN GENERAL MERCHANDISE.

Paid

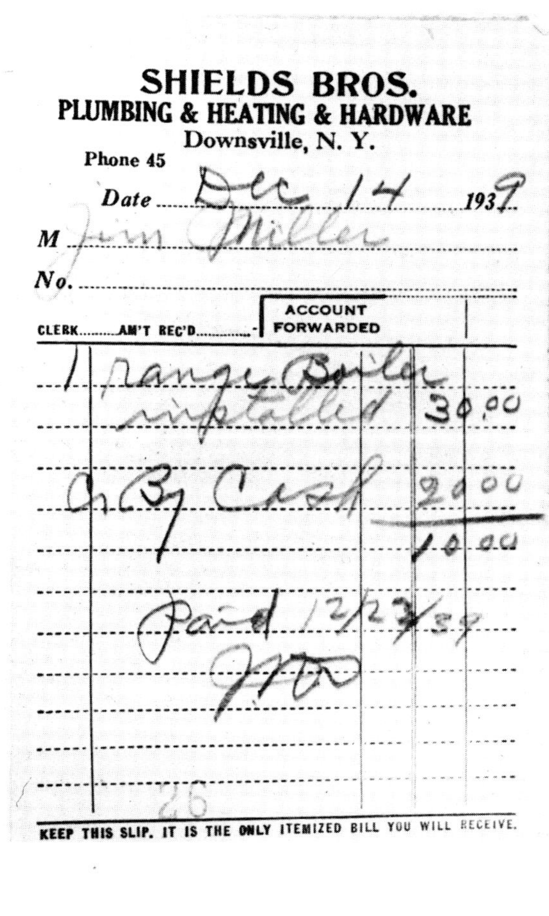

SHIELDS BROS.
PLUMBING & HEATING & HARDWARE
Downsville, N. Y.
Phone 45
Date Dec 14 1939
M Jim Miller
No.
CLERK ...... AM'T REC'D ...... ACCOUNT FORWARDED

1 range Boiler installed 30.00
By Cash 20.00
10.00

Paid 12/27/39

KEEP THIS SLIP. IT IS THE ONLY ITEMIZED BILL YOU WILL RECEIVE.

Covered Bridge at Shavertown.
Betty Berke Collection

SHAVERTOWN

Iron Bridge at Shavertown
The Ed Russell House is at the left in both photos.

Florence Russell Huggins writes that her father, Ed Russell, bought the house just across the river from the Presbyterian Church and next to the iron bridge over the East Branch of the Delaware River at Shavertown in 1922 or 1923 before the Presbyterian Manse was built next to the church on River Street. At the time Rev. and Mrs. Robertson were living in the house and he let them stay until the manse was ready. 37

Post Card View-Railroad Ave.,Shavertown,N.Y.-From left to right-
Dwight Shaver house,which was last owned by Scott Barnhart,Jim Bramley house, Railroad Depot, Armstrong Store, Uriah Sprague Store, and the Charlie Dutcher Hotel which was last owned by Carrie Elwood.
Mabel and Clara Miller Collection.

Post Card view of the interior of the Armstrong Store at Shavertown, N.Y. Kenneth Sprague Collection.
Some of the people are believed to be as follows from left to right; unknown, Ward Shaver, Levi Sprague, unknown child, unknown child, Uriah Sprague, Jake Atkin, unknown, unknown, Albert Armstrong wearing neck tie, the two sitting at lower right and barely visible are unknown.

Shavertown Water Company

After the railroad came to Shavertown in 1906, many business places and homes were built and there became a need for water for the village. The Shavertown Water Company came into existance in 1922. Water was brought to the village quite a distance from a reservoir situated up Shaver Hollow. Grant March has memories of this venture. In an effort to save money, a salesman pursuaded them to buy cheap pipe made from pieces of wood held together by band iron. It was not long before the wooden pipe began to leak in several places on its way to Shavertown.

Before this time the Corbett and Stewart Company had built an Acid Factory at Shavertown about 1916 or 1917. They used steel pipe to pipe water from an area that was dammed up on the Tremperskill stream on Hob Shaver's farm. The Acid Factory closed shortly after World War I. The Shavertown Water Company bought the used steel pipe from the Acid Factory and this gave them enough steel pipe to replace the wooden pipe.

At left is Bruce Armstrong with his feed and coal truck. In the background is a side view of Uriah Sprague's Store at Shavertown.

Dorothy Rice Collection.

The two photos on this page are of the Shavertown Fair. Jim Andrews Collection.

PROGRAMME

SHAVERTOWN FAIR,

SHAVERTOWN, N.Y.,

AUGUST 26, 27 AND 28,

1902.

TUESDAY, AUGUST 26.

9 00 a. m. to 3 p. m.—Receiving and arranging exhibits.

1 30 p. m.—Hayseed Race for single horses. Purse $6 00. First $3 00; second $2 00; third $1 00. Four or more to enter. Best three in five.

WEDNESDAY, AUGUST 27.

9 00 a. m.—Judging exhibits.

10 00 a. m.—Judging Special Premiums.

10 30 a. m.—Ladies' Driving, single.

11 00 a. m.—Hitch-up Race for single horses. Purse $5 00 First $3 00; second $2 00. Three or more to enter. Best two in three.

1 00 p. m.—Parade of Premium Stock.

2 00 p. m.—Farm Trot for single horses. Purse $10 00. First $5 00; second $3 00; third $2 00. Four or more to enter. Best two in three. No horse allowed in this race except those owned by farmers and used for ordinary farm work and never having trotted for money except in farm races.

2 30 p. m.—Double Farm Trot. Purse $10 00. First $5 00; second $3 00; third $2 00. Four or more to enter. Best two in three heats. No team allowed in this race except those owned by farmers and used for ordinary farm work and never having trotted for money except in farm races. No horse allowed in any farm race unless owned by the exhibitor at least 60 days before the fair.

3 30 p. m.—A Special Premium will be offered for best Decorated Rig, single or double. Purse $20 00. First $10 00; second $5 00; third $3 00; fourth $2 00. Entry fee in this exhibit $1 00.

THURSDAY, AUGUST 28.

9 00 a. m.—Judging Horses and Colts.

10 30 a. m.—Ladies' Driving, double.

11 00 a. m.—Ball Game between Fleischmanns and Downsville

1 00 p. m.—Race for Single Horses that have never trotted or paced better than three minutes. Purse $45 00 First $20; second $15 00; third $10 00. Four or more to enter. Best three in five.

1 15 p. m.—Trotting and Pacing Race for single horses; free for all. Purse $70 00. First $35 00; second $20 00; third $15 00. Four or more to enter. Best three in five.

☛ Application for entry in any race must be accompanied by 10 per cent of amount of purse. Any information will be given by Secretary.

JAS. BRAMLEY, President. L. E. HULBERT, Secretary.

RECORDER PRINT, ANDES, N. Y.

Shavertown Fair program at left.

Chip Sidorowicz Collection.

TURDAY, SEPTEMBER

SHAVERTOWN FAIR.

A Successful Exhibition—Good Displays in all Departments.

The third annual fair of the Shavertown Agricultural Society was held this week. The fair, which sprang from a few enthusiastic farmers and breeders coming together to show colts in the fall, has grown to be one of the best fairs in the county. The entries this year were far in excess of any previous year in all departments. President H. L. More and Secretary J. W. Davis have left no stone unturned to push the fair to the front.

Among the stock was the fine herd of Devons exhibited by E. T. Finch, of Granton. Other breeds were well represented.

G. H. Edgerton, of Franklin, shows a trio of Merino sheep and a collection of fine swine, consisting principally of Jersey Reds and Victorias.

Herman Wilson, of Wilson Hollow, had a pen of Cotswolds; also a bunch of wool sheared from a yearling ewe this season, which measures fifteen inches in length.

In the fruit and vegetable department was a fine display of potatoes, grain, grasses and the usual large pumpkin; besides apples, pears, plums, etc.

In the ladies' department is an elegant collection of flowers and fancy work.

The poultry department has outgrown the building, and one exhibitor shows over half his load from the wagon. The exhibit that is attracting the most attention in this department is the gold and silver Pheasants exhibited by Hillis Bros., of Davenport. The principal other exhibitors are Dennis Bros., of Bovina Centre, and Leonard Smith, of Bloomville.

The number of people on the ground was variously estimated at from 2,500 to 4,000, which is certainly a large crowd when we consider that all must come from a distance and in wagons.

Copy of a Newspaper article of the 1897 Shavertown Fair shown above.

These newspaper clippings of Shavertown news are believed to be from the Downsville News, as A.E. Peck was editor of that newspaper.

1903

SHAVERTOWN NEWS.

WHAT OUR REPORTERS FIND WORTHY OF NOTE.

Fresh Grist of Happenings Condensed and Boiled.

[From our own correspondent.]

Thomas Ford was in Delhi Monday on a business trip.

In conversation with one of the members of the Shavertown Agricultural Society this week he stated that it is very doubtful if another fair is held here. In his opinion he thought the society would disband, as it is nearly impossible to conduct a fair without state aid. A meeting will be held on Saturday evening of this week and definite action taken.

Several from this section will attend the social dance at the home of A. Franks, Union Grove, on Friday evening of this week.

Mrs. T. H. Tidd expects to join her husband at East Brookfield, Mass., in a short time. Mr. Tidd has a lucrative job in a shop at that place and, we hear, is "stuck on" the country and climate and prosperous times there.

Mr. and Mrs. James May of Delhi are at Thos. Ford's, where they expect to make their home the remainder of the winter.

George M. Atkins is quite ill at present. Dr. Brittain attends him.

At the fair meeting on Saturday evening of this week it may be decided to discontinue the fair for one year, a few of the members being in favor of this proposal, while others are not.

A social will be held at the home of Frank Norris on Tuesday evening of next week for the benefit of the U. P. church. All are invited to be present.

One of our citizens has established a news bureau and is contributing to several county papers at a "high" salary. He finds it much cheaper than to pay cash for the journals.

The late A. B. McCune, whose death occurred last Thursday, was an up-to-date farmer and dairyman and a progressive citizen, possessing many warm friends throughout the town of Andes. His family have the sympathy of all in their bereavement.

Some time ago there was a rumor afloat that a feed mill and store would be erected at this place. D. E. Boyd of Downsville and "June" Shaver of this place were spoken of as the head promoters of the enterprise, and all of us began to have visions of cheaper feed, without having to haul it a considerable distance. But alas! the dream is broken. Just now the above capitalists are silent on the subject, but they say that a railroad will revive them in an instant.

A. E. PECK, EDITOR.

SHAVERTOWN NEWS.

1903

WHAT OUR REPORTERS FIND WORTHY OF NOTE.

Fresh Grist of Happenings Condensed and Boiled.

[From our own correspondent.]

DEATH OF A. B. MCCUNE.

The death of A. B. McCune occurred last week Thursday at his home in Pleasant Valley, after an illness of about four months. The funeral was held at the house on Saturday, Rev. James Bruce officiating. Burial at Andes. Deceased was about 49 years of age and leaves a wife and three children, Mrs. Frank Wood and John McCune of Roscoe and Carl of this place. He was a brother-in law of R. D., Jennie and Mary Lamont of Delhi, who were present at the funeral. Deceased was born in the town of Bovina in 1853 and moved to Shavertown in 1878. He took possession of his present farm 13 years ago, and has been a member of the U. P. church about 16 years. At the beginning of his illness he was stricken with pneumonia and during his long sickness had a hard battle for life, suffering several relapses. He was an excellent citizen, a good friend and neighbor and will be greatly missed.

SECURED A NICE SUM.

The social held at the hospitable home of George Shaver last Friday evening for the benefit of O. E. Shaver, who has for a long time efficiently supplied the pulpit of the First Presbyterian church, was a grand success and largely attended. The sum of $42 was contributed.

OUR NEW BUTTER MAKER.

E. H. Truesdell of Potsdam, N. Y., has been engaged by Mr. Root to take charge of the creamery at this place the coming year. He is an excellent butter maker and will no doubt give complete satisfaction, although to please all of the patrons equally well is a big contract for anyone to undertake.

RICHARD WON FIRST.

As a result of the bean guessing contest at C. R. Gladstone's, Richard Mines and James Wilcox came off victorious, the former winning a handsome lamp and the latter a beautiful picture. The jar contained 11,077 beans. Miner's guess was 11,080 and Wilcox's 11,071. The next best guess, or 3d, proved to be a tie between Lester Hulbert, late secretary, etc., and Jas. Wilbur. The prize was the beans. The count was made and result announced on Saturday last.

ONLY ONE ANXIOUS.

There's nothing new in regard to the question whether a fair will be held at this place the coming summer or not. In case the society disbands and no effort is made to resuscitate it, it is quite evident that no fair will be held in this part of the East Branch valley in some time. While at Downsville recently on a feed buying tour we interviewed several of the business men and other citizens of that place and asked them why they did not take steps to have the fair at that place. Only one, a hotel man, seemed at all anxious for the transfer. The rest were indifferent.

OTHER MATTERS.

Among the recent real estate transfers in this vicinity we notice the following: Isaac Fletcher and wife to Dwight Shafer, $1,500, Dec. 1.

Mrs. Geo. J. Merritt of Brooklyn was a guest of her sister, Mrs. C. R. Gladstone, this week.

J. L. Reynolds of Shaver Hollow is desirous of selling his farm. Here is a chance for some man to invest on easy terms in a good farm.

The Downsville News, Jan. 2, 1902.
Florence Northrup Collection

A. E. PECK, EDITOR.

SHAVERTOWN TOPICS.

WHAT OUR REPORTERS FIND WORTHY OF NOTE.

Late Happenings Gleaned During the Past Week.

Rev. Jas. Norris of L. I., is visiting his mother and sister near this village. Rev. Mr. Norris is pastor of the 1st Presbyterian church of Glen Cove. In the early part of the present year his church presented him with a purse of $2,000 as a token of esteem.

Edwin Shaver is very sick with erysipelas.

There was a family gathering at A. Finkle's, Christmas day.

Mr. and Mrs. David Hoag spent Christmas with Mrs. Hoag's parents at Perch Lake.

W. W. Sheaver has hired to Archie Gladstone for the remainder of the winter.

A. Barnhart and D. W. Shafer are now preparing to build them each a boat for the next high water.

The Beech Hill school and the Lower Shavertown school held a Christmas tree in the school house at the former place Christmas eve. There was a large crowd and a fine display of presents.

Alfred Barnhart and D. W. Shafer have been compelled to move all their hay, owing to its getting wet by the recent high water.

Levi Shaver and Alex. C. Allen are building a new bridge across the Beech Hill stream by Mr. Shaver's house.

George B. Fletcher and family and Charles Gladstone had a very narrow escape from drowning one day last week while returning from Downsville. They had crossed the river, but were compelled to drive in it again near Mark Shaver's, when their wagon slipped into a deep hole formed by the water. It was with some difficulty that they managed to extricate themselves from their perilous position.

Dick and Dora Sprague Collection.

WED AMID CHRISTMAS CHEER

1913

K. Elizabeth Norris Became Wife of Thos Davis on Christmas Day

One of the social events of the year took place on Christmas day at high noon, when Miss K. Elizabeth Norris became the wife of Thomas Davis, the ceremony taking place at the home of the bride's parents, Mr. and Mrs. F. D Norris, Shavertown, N. Y.

The ceremony was performed by Rev Lincoln R. Long, of Margaretuille, in the presence of about 40 relatives and friends of the bride and groom, and was one of unusual prettiness. Promptly at the appointed hour the bridal party entered the parlor to the strains of Mendelssohn's wedding march played by Miss Mabel Hulbert, and approached the evergreen arch which was tastily trimmed with snow white garlands and wedding bells, beneath which the solemn and impressive words were repeated that made the two man and wife.

The bridal party was led by six girl friends of the bride, Misses Edna, Letha and Jennie Hulbert, Carrie Sprague, Janie Shaver and Mary Davis, and as they neared the arch they separated and roped off the aisle with pure white garlands through which the bridal party were to pass. First came the little flower bearers, Hazel Davis and Master Conger Norris, carrying the rings on satin cushions. Following them came Miss Gertrude Norris, maid of honor, and Floyd Davis as best man; the groom next and on his ar[illegible] the fair bride, who was attired in [illegible] dress of white silk embroidered net over white messaline with crystal an[illegible] seed pearl trimming, and wearin[illegible] white lilacs. The six girls were dres[illegible]ed alike in white chiffon and shad[illegible] lace with pink rose bud trimmin[illegible] The maid of honor wore white emb[illegible]dered lace over light blue. Miss Ma[illegible] Hulbert wore a dress of pink cha[illegible]meuse with blue forget-me-not an[illegible] rose buds.

Following the ceremony and con[illegible]gratulations a wedding repast was ser[illegible]ed and Mr. and Mrs. Davis departe[illegible] amid a shower of rice and old shoes t[illegible] take the 3:20 train for a short weddin[illegible] trip which will include Buffalo, Nia[illegible]ara Falls and some parts of Canada.

The bride was the recipient of ma[illegible] beautiful presents, including s[illegible] silver, cut glass, 156-piece set Chi[illegible] dishes, table linen, etc. A Guest.

The two photos above were taken by George Hoag as he delivered mail along the Dingle Hill Road from Shavertown in the year of 1918.

The ticket at left comes from the days when Silent Movies were shown in Fletcher Hall which was located in back of the Shavertown Store and Post Office. The silent movies were shown by George Hoag in 1930.

Copy of a post card dated 1936 at left shows a bobcat shot by Grant March. It measured 53 inches long and weighed 36 pounds. The house in background was situated in back of the Shavertown Store and Post Office. The store and post office was formerly owned and operated by Lester Hulbert and his wife, Amanda Hulbert (Fletcher) and for many years after that time until construction of the Pepacton Reservoir the store and post office was operated by Jake and Inez Atkin. The house was Shavertown's oldest house and it was last owned by Woodrow Dent.
Daryl Atkin made and sold post card views of this bobcat at Atkin's Store.

In the photo at lower left is another bobcat which got caught in a trap along the area of the Mary Smith Road, town of Andes. The bobcat had to be destroyed.

Both views are from Grant March Collection.

Lester Hulbert's business card is shown at right below.

L. E. Hulbert

General Merchandise

Also Postmaster

Shavertown, N. Y.

A Century of Service

AUGUST 26, 1851 - AUGUST 26, 1951

100th ANNIVERSARY

of

First Presbyterian Church

of Shavertown

ROBERT D. MAHAFFEY, Pastor

OFFICERS

TRUSTEES

Bruce Armstrong, Chairman

Maude Cox — Ben Conklin

George Hoag — Leland Titch

ELDERS

Scott L. Barnhart — Ray Bramley

Milford Butler — George Rinehart

George D. Scudder — Edwin Finkle

Benjamin Conklin — Leland Titch

Mrs. Ada Mayer, Organist

Mrs. Hazel Finkle, Assistant

SUNDAY SCHOOL

Leland Titch, Organist and Superintendent

Mrs. Ellen Van Steenburg, Secretary and Treasurer

Mrs. Marian Armstrong, Teacher

Mrs. Doris Smith, Teacher

MORNING WORSHIP

Joint Shavertown and Downsville Congregations

Begins 11 A. M.

Conducted by Rev. Millard B. McCune

ORDER OF WORSHIP

Prelude
Doxology
Invocation and The Lord's Prayer
Hymn 140 . . . "The Church's One Foundation"
The Apostles Creed
The Scripture . . . . Romans 4
Prayer
Special Music
Announcements
The Offering
Hymn 149 . . . "The Church in the Wildwood"
Sermon . . . . "Our Faith, Then and Now"
Rev. Millard B. McCune
Hymn 219 . . . . "Faith of Our Fathers"
Benediction

Free Lunch Served at Noon

AFTERNOON SESSION - 2 P. M.

Music and Singing
History of Church . . . . . Arthur Russell
Introduction of Visiting Pastors
Closing Services

EVENING SERVICE - 8 P. M.

By the Rev. Emily Danneheimer of the Cabin Hill Church

ORDER OF WORSHIP

Prelude
Hymn 435 . . . . . . "My Savior's Love"
Hymn 231 . . . . . . "Under His Wings"
Responsive Reading . . . . Selection 551
Choir Selection . . "Shall We Gather at the River"
Scripture Reading
Prayer
Announcements
Presentation of Offering
Hymn 257 . . . . . "Nearer, Still Nearer"
Vocal Duet . . . . . . "Give My Heart"
Miss Danneheimer and Miss Gage
Sermon . . . . Miss Danneheimer
Closing Hymn 288 . . . . . "Close to Thee"
Benediction

The Advent Church-

At left is a copy of a news clipping from the collection of Mabel and Clara Miller.
"This church at Shavertown was awarded $12,500 by a commission of appraisers after it was condemned by New York, because of the Pepacton Reservoir."

In the photo at the right, they are starting to construct Uriah Sprague's store at Shavertown.
Below is a photo of Uriah Sprague's store after it was built.
Both photos are from the collection of Dorothy Rice.

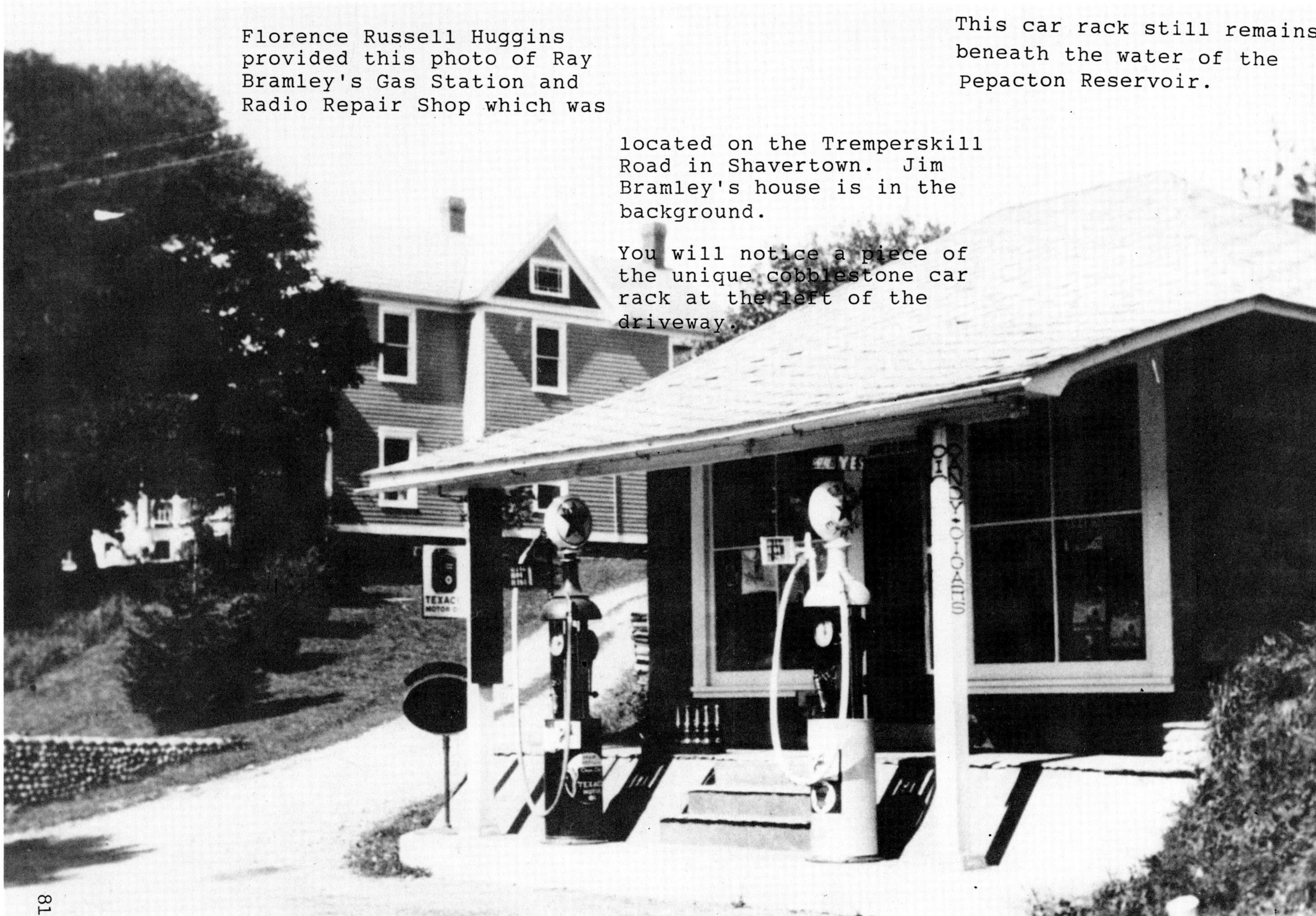

Florence Russell Huggins provided this photo of Ray Bramley's Gas Station and Radio Repair Shop which was located on the Tremperskill Road in Shavertown. Jim Bramley's house is in the background.

You will notice a piece of the unique cobblestone car rack at the left of the driveway.

This car rack still remains beneath the water of the Pepacton Reservoir.

REPAIR SERVICE TUBES TESTED FREE

EXPERT RADIO SERVICE

RAY BRAMLEY

CROSLEY RADIOS

RAYTHEON TUBES

SHAVERTOWN, -:- -:- NEW YORK

This copy of Ray Bramley's business card is from the collection of Grant March.

ELECTRIC LOVE

If she wants a date—Meter.
If she comes to call—Receiver.
If she wants an escort—Conductor.
If you think she is picking your pocket—Detector.
If she's slow of comprehension—Accelerator.
If she goes up in the air—Condenser.
If she's hungry—Feeder.
If she's a poor cook—Discharger.
If she eats to much—Rectifier.
If her hands are cold—Heater.
If she fumes and sputters—Insulator.
If she wants a holiday—Transmitter.
If she is narrow in her views—Amplifier.

At left is Ray Bramley standing on the unique cobblestone car rack next to his gas station and radio repair shop on the Tremperskill Road at Shavertown. Photo is from the collection of Dorothy Rice.

The oldest house in Shavertown at the time of the construction of the Pepacton Reservoir is shown in the photo above. It was last owned by Woodrow Dent.

Both photos are from the collection of Dorothy Rice.

The Armstrong house was the newest house in Shavertown at the time of the construction of the Pepacton Reservoir.

UNION GROVE

The photo of Union Grove above
is copied from "Delaware County,
N.Y. History of the Century 1797-
1897 Centennial Celebration
June 9 & 10, 1897".

Edited by David Murray, LL.D.
William Clark, Publisher,
Delhi, N.Y. 1898, page 279.

This church at Union Grove settled with New York, without going before a commission, and received a check of $4,750. There are no real estate appraisers in The News office. But to the common

Congregation of the Union Grove Methodist church, a few hundred feet away, allowed its case to go before a commission. Last week the commission awarded $15,500 to the Methodists for their building. The buildings appear to be of comparable quality outside. The commission method seems to bring higher awards to valley folks than a direct sale.

Both photos of the two churches at Union Grove are copies of newspaper items from the collection of Clara Miller.

Photos above are copies of a newspaper collection of Mabel and Clara Miller.

This mill was located just outside Union Grove along the Barkaboon Stream.

"Relic of another day, the old Union Grove mill stands empty and idle beside a tumbling waterfall on the Barkaboon stream. Built in 1883, it ground most of the flour and feed for people of the section in a day when everything was home related and home made.

In the photo above at right, "One of the three sets of grinding stones in the Union Grove mill. They were brought from France in 1883."

Below is a post card view of the bridge at Mill and Main St., Union Grove.

Jenkin's Falls, Union Grove
above is a post card view
from the collection of
Dorothy Andrews.

Below is a post card view
of Main Street, Union Grove.

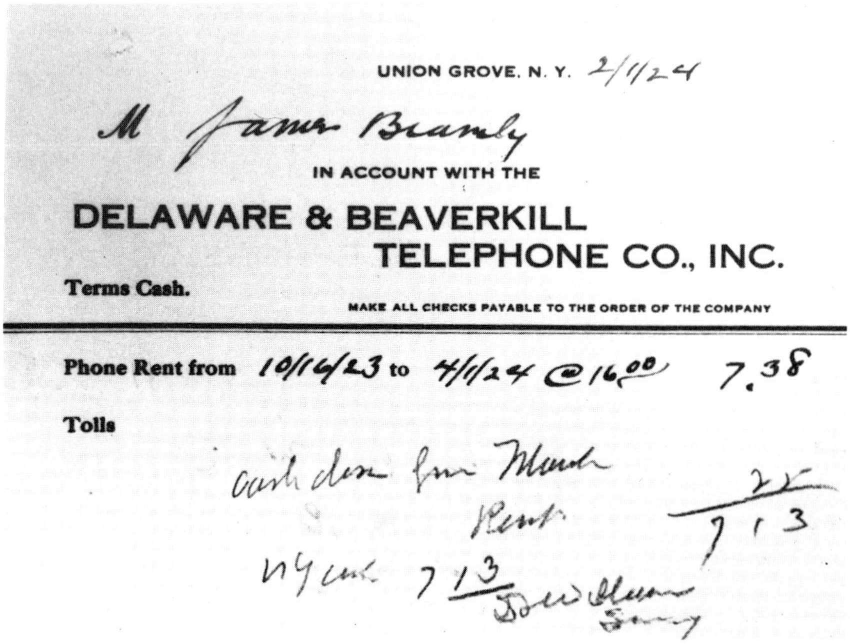

UNION GROVE, N. Y. 2/1/24

IN ACCOUNT WITH THE

DELAWARE & BEAVERKILL TELEPHONE CO., INC.

Terms Cash.

MAKE ALL CHECKS PAYABLE TO THE ORDER OF THE COMPANY

Phone Rent from 10/16/23 to 4/1/24 @ 16.00 7.38

Tolls

The bill from the Delaware & Beaverkill Telephone Co., Inc. and the label from Union Grove Mills are from the collection of Dorothy Andrews.

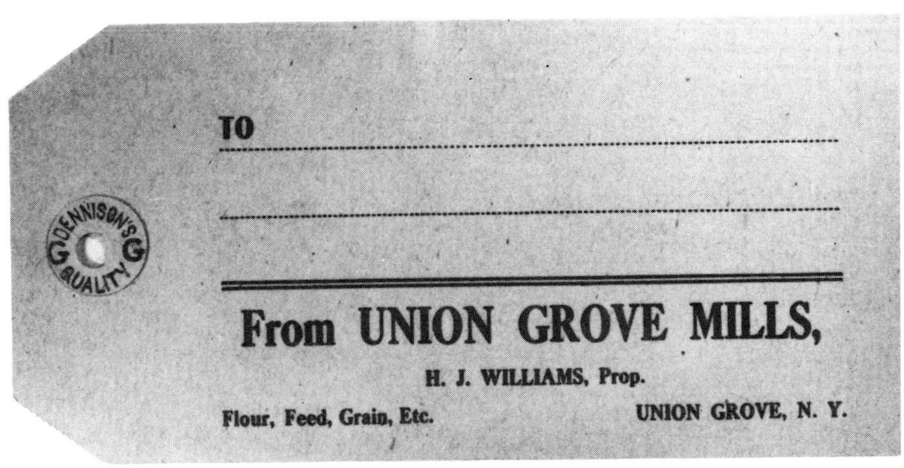

ARENA

The photo above shows the Iron Bridge connecting Arena with N.Y.S. Route 30. The village of Arena is in the background.
Mabel and Clara Miller Collection.

From the Catskill Mountain News 'Yellowed Pages'. "While digging a grave in Arena, workmen found a skeleton of a very large man. The man had been burried in a hemlock coffin and was dressed in buckskin breeches." (March 6, 1914)[38]

The following newspaper clipping was sent to me from Mrs. Olive Wilson.

OCCUPANT BORN IN 1711 , CAPTAIN IN REVOLUTION

"The oldest marked grave in the Pepacton Reservoir area, according to "Uncle" Don Henry, B.W.S. employee, in charge of cemetery removals, was found in the "old" Rural Arena cemetery by means of a 2.8 foot, by 1.5 foot, 1 1/8 inch slab with the following inscription:

In Memory of
Capt. Samuel Dunham
who departed
this Life
July 27th, 1792
AE 81

"According to Mr. Henry, "every stone in the four major cemeteries and several small private plots of the area has been measured, and the inscription copied."

"The four "large" cemeteries include the one at Shavertown containing 733 bodies, one at Union Grove containing 217, two at Arena, the "old", or "rural" one containing 847 bodies and the "new" or "village" cemetery containing 315.

"The minor, or smaller cemeteries in the area are the Edget plot for the Byington farm with 183 graves, the Shaver plot at the mouth of Cole's Clove with 26, the Flynn plot between Pepacton and Shavertown with 15 and the Cat Hollow plot at the mouth of Cat Hollow.

"The latter was formerly known as the "darky burying ground" according to legend, because of the fact that in the 1700's a wealthy slave owner by the name of Cole settled in the mouth of the Clove which bears his name and started a burying ground for his slaves, who also took the Cole name, in Cat Hollow.

"In all, approximately 3,000 bodies are to be removed from the reservoir area before water flows to the dam." 39

"The old Arena gristmill is one of the next to go. To be sacrificed for Pepacton reservoir, it is still sound of timber, could be operational again in a short time."

Below- "This wooden cogwheel served for years in the Arena mill. Mechanism is of wood in beautiful workmanship."

Both photos are copies of newspaper clippings.

Copy of a ink blotter shown below.

Photo of the Advent Christian Church, Arena, N.Y. at right comes from the Collection of Mabel & Clara Miller.

At left is a copy of a post card of Maple Ave. looking north, Arena, N.Y. Chip Sidorowicz Collection.

At right is a post card copy of J.D. Murdock's Residence, Arena, N.Y. Jim Andrews Collection.

Copy of a matchbook cover.

Chip Sidorowicz Collection.

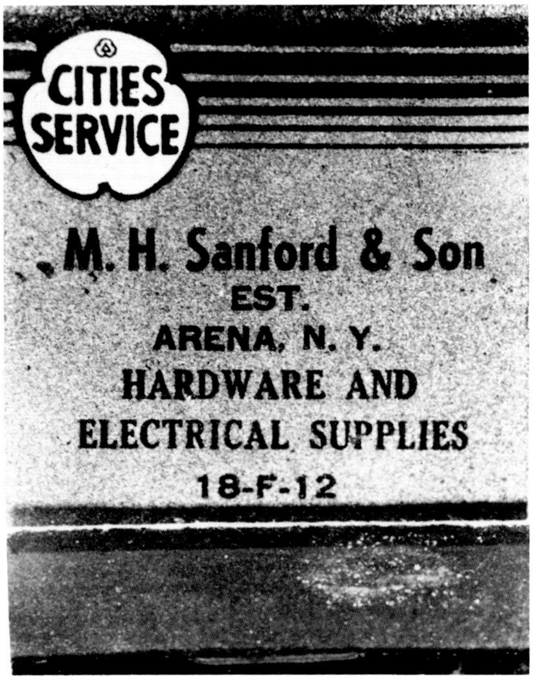

Copy of a fan below from the General Merchandise Business of Mrs. Katharyn Dickson, Arena is from the Chip Sidorowicz collection.

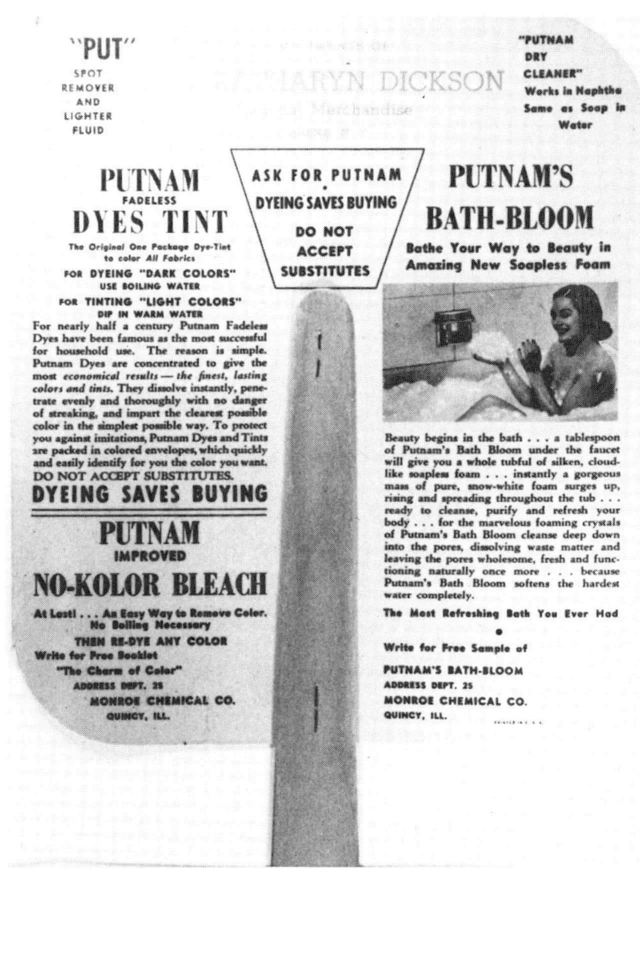

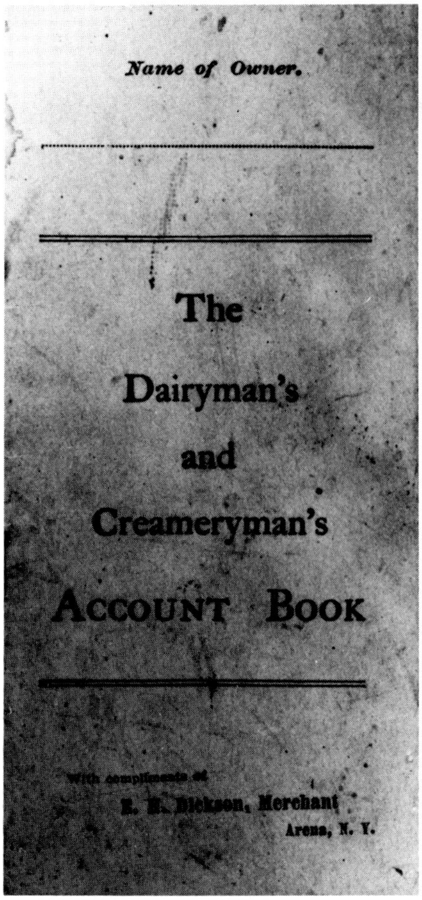

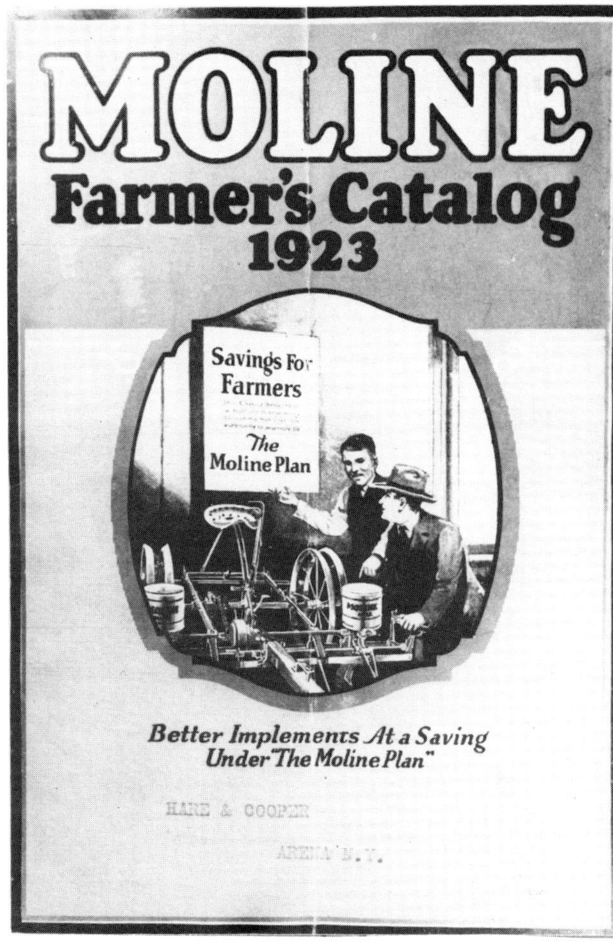

The account book from E.H. Dickson, Arena and the Catalog from Hare & Cooper, Arena are from the Chip Sidorowicz Collection.

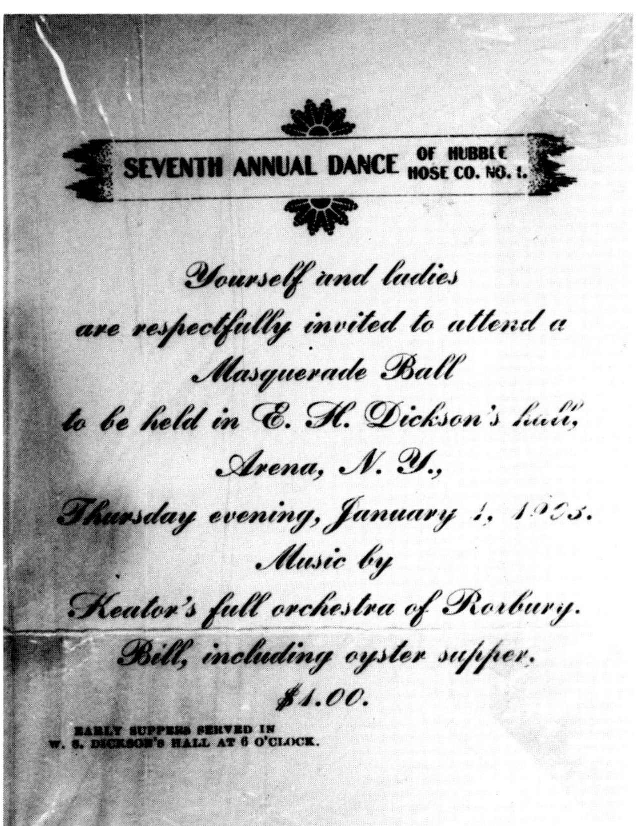

The Seventh Annual Dance invitation at left comes from the collection of Dorothy Andrews.

Copy of a Post Card
at left.

At right - V-C Fertilizers

for sale by Holmes &
Bussy - Chip Sidorowicz
Collection.

Post Card view of Old Mill,
Arena, is from the collection
of Chip Sidorowicz.

The Downsville News
Nov. 2, 1916
Florence Northrup
Collection.

THE NEWS IN ARENA.

THE JACKSONBURGH BRIDGE—REMAINS BROUGHT TO ARENA.

The remains of Will More of Roscoe were brought to Arena last Friday for interment. Deceased was about 63 years of age and is survived by a wife and three children. He was a former resident of Arena.

On Wednesday night of last week a large auto truck loaded with furniture broke through the old covered bridge at Jacksonburgh, below this village. The bridge was quite badly damaged, as was also the truck, which was loaded with furniture enroute from New York city to Stamford, via Liberty. This is not the first mishap of the kind in the town of Andes and it looks as though the bridges were not build to withstand heavy traffic. It is understood that the Andes town board wants to close the above bridge, instead of repairing it or buying a new one, but the proposition will be fought bitterly by the residents of Union Grove, as the highway from the bridge to Arena on the Union Grove side of the river is in the town of Middletown and is in a very shocking condition.

Ward Bryant of Union Grove, who is as well known in Arena as he is at home, is now raising Guinea pigs which he sells to laboratories for inoculating purposes. He is also raising Belgian hares for the market. Ward says that for delicious eating qualities the Belgian hare has everything stopped for a mile. P. S. He has gone out of the ginseng business, for the reason that so many people break through and steal.

S. W. Brown is still booking orders for lawn swings, and all kinds of turned goods. He did a large business in this line before his mill at Pepacton was burned to the ground. But of late he has not been operating his mill as, in a recent accident, he sustained a broken rib and other injuries, pre-

Stan's Tavern, owned and operated by Stanley Sidorowicz, was situated along Route 30, above Arena. It was first a gas station as shown in the photo above. It was known as Stan's Tavern for many years until the Pepacton Reservoir was built.

Photos come from Chip Sidorowicz Collection.

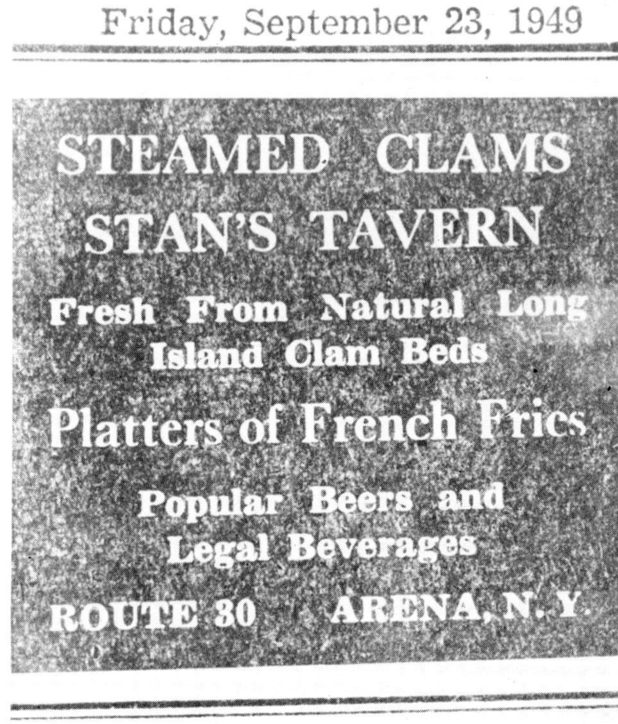

Above is a view of the interior of Stan's Tavern. At right is a copy of an advertisement and below is a copy of his business card.

All 3 items are from the collection of Chip Sidorowicz.

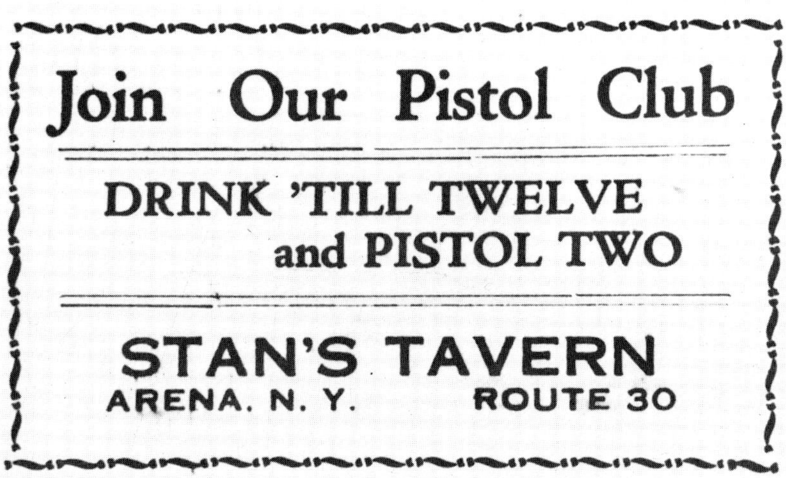

At left - Post Card View of Arch Bridge near Arena is from the collection of Jim Andrews.

Photo Copy of a Newspaper Clipping below

"In the picture just above you can see another old fireplace. It was part of the McLean log cabin at Dunraven. Standing beside it is Mrs. Eliza Kathman, the last owner before the property was bought by New York city Board of Water Supply and torn down to make way for the Pepacton Reservoir."

This fireplace located up above Arena near Dunraven had stood as a landmark for many years. My Great-Grandmother, Mary McLean Fenton had lived there when young.[40]

THE ANDES FOUNDRY

T. S. MILLER, Proprietor.

MLLER'S REVERSIBLE FLATLAND PLOW,

Warranted to Give Entire Satisfaction

Chilled Iron Plows.

These items are copied from The Andes Recorder
Andes, Delaware County, N.Y., Thursday
July 11, 1889.

GRAND CELEBRATION!

AFTER WE CELEBRATE THE GLORIOUS FOURTH.
AFTER WE HAVE SEEN THE FAT MAN'S RACE.
AFTER WE HAVE EATEN THE CLAMS, ETC.
AFTER ALL IS QUIET AGAIN THEN COMES

HAYING.

We think of our machinery, tools, etc., and do not forget we must also have something to eat. Now we have been preparing for this and have on hand an extra large supply of

PROVISIONS, FLUR, FISH, RICE, TEAS, COFFEES, SUGARS, ETC., ETC.
HATS, SHOES. SUMMER PANTS, OVERALLS, ETC.

Also a complete stock of

LADIES SHOES, PARASOLS, GLOVES, MITTS,

and all Summer Goods at extremely low prices.

The highest market price paid for WOOL, BUTTER, EGGS, MAPLE SUGAR and other produce at

THE BIG STORE ON THE CORNER

J. W. DICKSON.

LIVERY

SALE, EXCHANGE AND BOARDING

STABLE

The best Horses and the finest carriages ot let both day and night. Having enlarged mv barn, I heve now one of the finest stables in Central New York, and am prepared to furnish

THE BEST RIGS

Outside of the large cities (from Clark's Factory to the "Hook"

ON REASONABLE TERMS.

With thanks for liberal patronage in the past, I hope by fair treatment to merit a continuance of the same in the future. Remember that the

UPTOWN LIVERY STABLE

is the place to go.

All parties wishing to Buy, Sell or Exchange horses are invited to call.

D. A. ARMSTRONG.

FRESH GOODS!!

Added to Our Stock Every Week.

SOME SPECIALTIES

IN LADIES' DRESS GOODS

we have some special bargains and can furnish anything in the line at prices that are as attractive as the goods.

GENTS', YOUTHS' AND BOYS' CLOTHING

is also a special feature of our extensive stock. We don't ask anyone to take our word for it, but respectfully invite all who need anything in that line to examine our stock, we've got the goods and "don't you forget it."

WHITE GOODS, HAMBURGHS, ETC.

It will be found to be a pleasure to select from our stock in these lines, and see completeness.

HATS AND CAPS

The new and nobby; the latest and best. Styles for everybody. We've got 'em.

Also a fine line of

HONESDALE BOOTS AND SHOES.

WALL PAPER

Now we can promise you a large stock to select from and prices lower than ever before.

FERTILIZERS

You want something in this line and we shall make it an object to deal with us.

LUMBER

Any one needing hemlock lumber of any description will do well to call and get our prices. First-class stock on hand and cut to order. Also large quantity of first class hemlock shingles and lath.

D. A. FLETCHER, ARENA, N. Y.

(The above ad comes from The Andes Recorder, Andes, Delaware County, N. Y., Thursday, July 11, 1889. Vol. 22, No. 28)

The subscription rate of this newspaper was $1.00 a year.[41]

Copies are from July 11, 1889 issue of the Andes Recorder.

U. & D. RAILROAD.

WEST.			EAST.	
Hobart Exp	Steambt Exp	STATIONS.	N. Y. Exp	Steambt Exp
Daily ex. Sun.	Daily		Daily ex. Sun.	Daily
P. M.	A. M.		A. M.	P. M.
2 35	7 40	..RONDOUT..	11 40	5 15
4 12	8 49	..PHŒNICIA..	10 35	4 12
4 57	9 36	..PINE HILL..	9 57	3 42
5 12	9 50	.GR. HOTEL..	9 50	3 35
5 20	9 59	GRIFFIN COR	9 30	3 19
5 32	10 11	..ARKVILLE..	9 18	3 08
5 40	10 19	KELLEY COR.	9 06	2 58
5 45	10 24	HALCOTVILL'	9 01	2 53
5 58	10 38	..ROXBURY..	8 47	2 40
6 12	10 53	..GR. GORGE.	8 32	2 25
6 31	11 12	..STAMFORD..	8 10	2 05
6 40	11 20	...HOBART...	8 00	1 55
P. M.	A. M.		A. M.	P. M.

Other trains leave Arkville, GOING EAST,—Arkville Express at 4:50, a. m., except Sunday, on Monday at 5:50, a. m. Steamboat Express on Sunday at 8:36, p. m. GOING WEST, arrive, Arkville Express 9:05, p. m. Hobart Express at 7, p. m.

ARKVILLE & DELHI STAGE LINE

VIA ANDES. SUMMER SCHEDULE.

Going East—Leaves Delhi at 8 a. m., Andes at 11 a. m.—connecting with train at Arkville for N. Y. Going West—Leaves Arkville at 11 a. m., after arrival of N. Y. train, Andes at 2 p. m. Ar. Delhi 5 p. m Through tickets for New York for sale at Delhi and and Andes. J. J. COWAN, Proprietor.

N. Y. ONTARIO & WESTERN R. R.

DELHI STATION.

Arrive—7:00 a.m., 3:50 and 9:00 p. m. Leave—9:25 and 10:40 a. m.

RONDOUT BOATS—NIGHT LINE.

RONDOUT TO NEW YORK.

Leave Rondout—at 6 p. m.—"City of Kingston" on Mondays, Wednesdays and Fridays; "J. W. Baldwin" on Sundays, Tuesdays and Thursdays. Returning, leave New York at 4:00 p. m. On Saturday City of Kingston leaves New York at 1 p. m.

THE ANDES MUTUAL FIRE INSURANCE CO.

ANDES, N. Y.

OFFICERS—William. S. Doig, President; H, N. Davis, Vice President; Wm. C. Laing, Secretary; Isaac Samuels, Treasurer.

DIRECTORS—John E. Gladstone, Isaac Samuels, W. Oliver, Adam Sutherland, Jas. Brotherton, Geo. Liddle, Wm. S. Doig, Jas. Fletcher, H, N. Davis, Alex. McQueen, Jas. Jack, D. Muir, J. G. King.

Painting Lessons.

Mrs. M. L. Fuller of Delhi wishes to announce that she is organizing a class in oil painting at Andes to begin Sept. 1, next. Any who may wish to join can learn further particulars at the RECORDER office.

For Sale!

One of the best pair of oxen in Delaware county. Will mow rake or haul hay without leaving load, or your seat on rake or mower. Will plow as much as horses. Also yearling colt. Inquire of C. A. HUNT, Lumberville, Del. Co. N. Y.

Jersey Calves.

The subscriber has six high-grade Jersey calves for sale. ROBERT BIGGAR, Coulter Brook, Bovina. 28-w2.

Gerry Camp news item comes from The Andes Recorder, July 3,1914.

GERRY CAMP FOR BOYS OPENED

Eighty-two Boys Arrived Wednesday at Camp Maintained by R. L. Gerry

The Boys Camp maintained by Robert L. Gerry on the Wight farm in southern Bovina, opened Wednesday and will remain open for two months.

Wednesday evening 82 boys arrived at Andes and were conveyed to the Camp. They will remain for two weeks and then a new group will take their place. Any whose health may require it will remain more than the two weeks.

The Camp officers are: Governor James L. Whitcomb; Captains Donald H. Morse, Gregory Prout and Linn Bruce jr. The chaplain is the Rev. Phoenix.

Farming

In the book "Three Centuries in Delaware County N.Y." published by New York Swiss Alps of Delaware County, 1933, John F. DeVine writes, "Throughout the one hundred and seventy-one years that Delaware County has been the home of white men it has been primarily a dairying section"... "Delaware County is known as one of the most important milk-producing centers in the United States."

Even barbed wire which is commonly used for fences to restrain cattle is locally conected. It was invented by "Hiram" Benjamin Scutt who was born in Andes in 1842, the son of "Hy" Scutt. "Benjamin went to Illinois in 1860. After serving in the Civil War he settled at Joliet, where he conceived the idea of building fences of a new type of wire that would make trespassing uncomfortable, if not decidly risky."

"In 1891 the Dairymen's Association took, in Bovina what is believed to have been the first cattle census, showing that one hundred and seventeen farmers in that township owned 2,668 cows or an average of twenty-three each."...

"The last cattle census taken by Delaware County Farm Bureau shows there are 100,381 dairy cattle 70,000 of which are milking age on 3,957 farms. This would give an average of 26 cattle to a farmer."[42]

Ethel Bussy Coulter mentions, "A tub of Delaware County butter was judged the best in the world at the Great London Exposition in 1851. Another sample taken from the Close farm near Arena not only took first prize at the World's Fair at Chicago in 1892 but also received a perfect score of 100. This mark had never been reached before by any butter maker at a World Show."[43]

Memories of David Jacobson

By Bob Jacobson

The dairy farms in the early 1900's were small and for extra income the farmers cut wood and sold it to home owners to heat their homes during the winter months.

Hardwood trees were cut and piled with the help of neighbors for a few days. It was cut to stove length with a buzz saw which was run with a gasoline engine. On David's farm it was a five horse power International. Sometimes the pile of stove wood totaled one hundred cords.

In the winter after the dairy farm chores were done for the morning David would load three cords of wood. If there was enough snow it was loaded on a sleigh, otherwise it was loaded on a wagon. He then started from his Holiday Brook farm for Downsville.

One of the teams that David had made many such trips with was a pair of Mustang horses named Prince and Kit. These horses were branded and came here from the west. Horses were brought by the Ontario and Western Railroad to Walton. These were wild horses and at Walton they were herded over Walton Mountain to Downsville where they were sold at auction near the site of the former Eagle Hotel. Prince and Kit weighed 2190 lbs. with the harness on. The wood weighed about 1800 lbs. per cord. The bob sleighs added more weight.

It was two miles to Pepacton. The covered bridge at Pepacton crossing the East Branch of the Delaware was a very hard pull for a pair of horses pulling sleighs through it. It was 168 feet in length.

Sometimes teamsters got stuck in the bridge and a extra team of horses was hooked on to move these heavy loads on through the bridge. In later years a truck might be used to help out. Once they got through the bridge it was another six miles to Downsville. If it was cold, David walked to stay warm otherwise he rode on the load of wood. At Downsville the wood was unloaded, the money was collected ($4.00 per cord) and then he traveled the eight miles back to the Holiday Brook farm.

After the horses were unharnessed, put in the stable, fed grain, hay and water, it was time to go back to the cows and all of the evening dairy farm chores. Finally, it was the end of a long winter's day for a tired farmer.

One time he delivered a load of wood at Downsville for a lady. She was very upset because he had hauled in three cords of wood when just the day before a teamster had gotten stuck with one ton of coal. She threatened to have him arrested, but no whip had been used on these horses. David was just a little smarter, he used a better route getting in the driveway! By the time the wood was unloaded the lady had cooled off and it was back to business as usual![44]

It is interesting to note the following from "Three Centuries in Delaware County N.Y." - "As Dairying always has been the principal industry in Delaware County, it undoubtedly always will be, because of the climate, soil and topography of the area."[45]

The Pepacton and Cannonsville Reservoirs acquired a lot of the river farms and since these reservoirs were built, a large number of dairy farms have continuously faded into nonexistence. A large part of Delaware County is presently owned by people from the Metropolitan area and these places are now second and seasonal homes. What was once beautiful green meadow land and pasture for grazing cattle is now fast growing into brush land and much sub-dividing of farm land has taken place as the number of second homes continue to grow.

Winter scene at the Byron Huggins farm.
This farm was located along the back river road and just across the East Branch of the Delaware River from Shavertown.

Josephine Huggins Noonan Collection.

Mabel and Clara Miller Collection.

In earlier days George Miller used to go around to area farms with the horse and wagon shown in photo to white wash barns. Notice the hand pump on the wagon.

George Miller was a cousin of another area resident - George Robert Miller,

It is believed that the above photo was taken on the former Bill Bouw farm on Holiday Brook.

The photo at left shows Jim Miller at work at the Pepacton Feed Store.

Mabel and Clara Miller Collection.

This photo was taken when the feed store was owned and operated by Ed Green. Feed came in box cars on the train and it had to be bagged in 100 lb. hemp bags. They also had white and wheat flour in bags and some of it was in barrels. They also had buckwheat flour.

In later years the feed store was owned and operated by Rube Johnson until the Pepacton Reservoir was built.

The photo above shows Ed Green with the ponies in front of his house. This was first a hotel and there was another house on the farm where the Green family lived. The hotel was operated by Zeke Tiffany. Mary Deighton Flynn and Josh Hafele worked at the hotel. After Rube Johnson bought the farm, the house where the Green family lived was torn down. Rube Johnson owned the farm with the house shown in the above photo until the Pepacton Reservoir was built.

Photo comes from the collection of Mabel and Clara Miller.

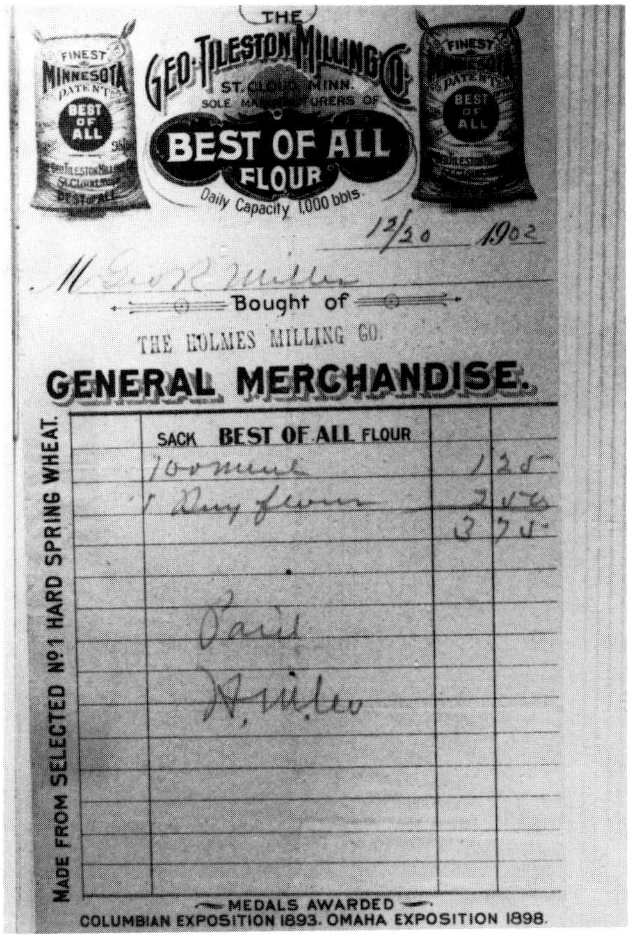
THE GEO. TILESTON MILLING CO.
ST. CLOUD, MINN.
SOLE MANUFACTURERS OF
BEST OF ALL FLOUR
Daily Capacity 1,000 bbls.

FINEST MINNESOTA PATENTS BEST OF ALL

12/20 1902

M Geo R Miller

Bought of

THE HOLMES MILLING CO.

GENERAL MERCHANDISE.

MADE FROM SELECTED No 1 HARD SPRING WHEAT.

SACK BEST OF ALL FLOUR	
100 meal	1 25
1 Buy flour	2 50
	3 75

Paid
H. Miller

MEDALS AWARDED
COLUMBIAN EXPOSITION 1893. OMAHA EXPOSITION 1898.

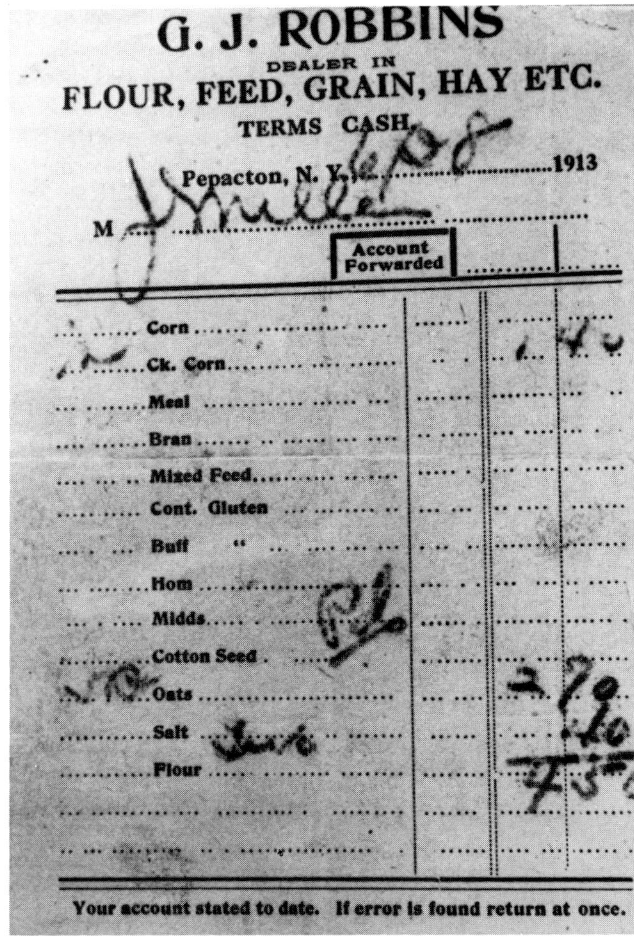
G. J. ROBBINS
DEALER IN
FLOUR, FEED, GRAIN, HAY ETC.
TERMS CASH

Pepacton, N. Y. 1913

M ....

Account Forwarded

Corn		
Ck. Corn		1 40
Meal		
Bran		
Mixed Feed		
Cont. Gluten		
Buff "		
Hom		
Midds		
Cotton Seed		
Oats		2 90
Salt		1 10
Flour		4 50

Your account stated to date. If error is found return at once.

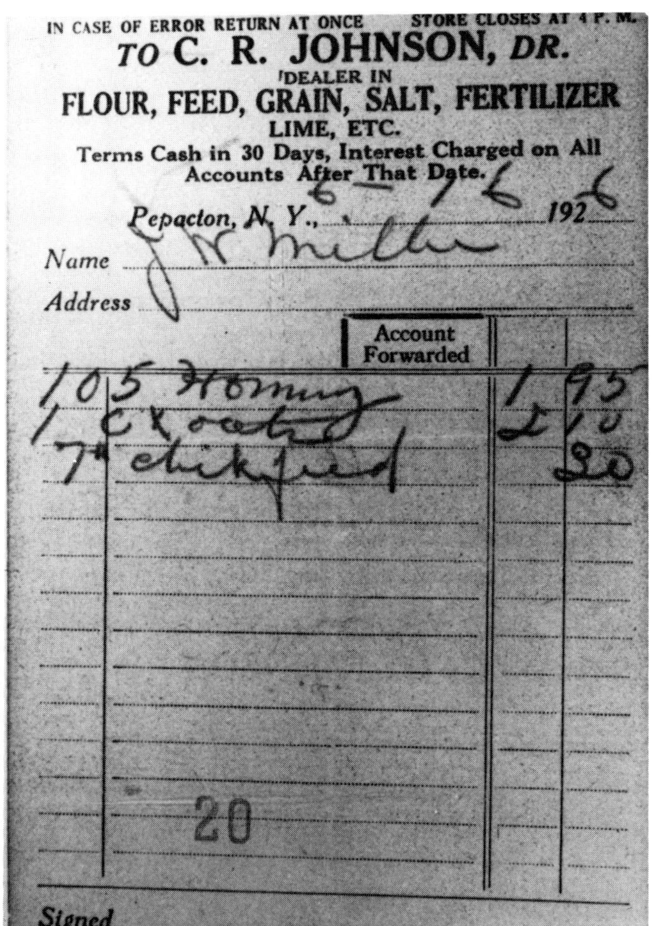
IN CASE OF ERROR RETURN AT ONCE STORE CLOSES AT 4 P. M.

TO C. R. JOHNSON, DR.
DEALER IN
FLOUR, FEED, GRAIN, SALT, FERTILIZER
LIME, ETC.
Terms Cash in 30 Days, Interest Charged on All Accounts After That Date.

Pepacton, N. Y., 6-16 1926

Name J R Miller

Address

Account Forwarded

105 Hominy	1 95
1 Ck oats	2 10
7# chikfeed	20

20

Signed

Copies come from the Mabel and Clara Miller Collection.

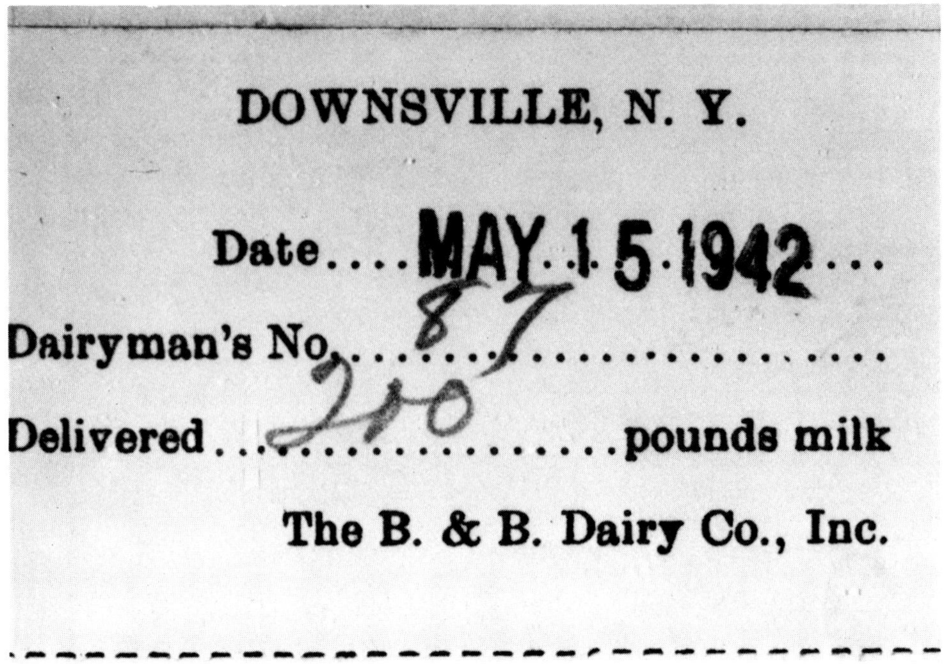

DOWNSVILLE, N. Y.

Date....MAY 15 1942...

Dairyman's No. 87

Delivered 210 pounds milk

The B. & B. Dairy Co., Inc.

Copies come from the collection of Mabel and Clara Miller.

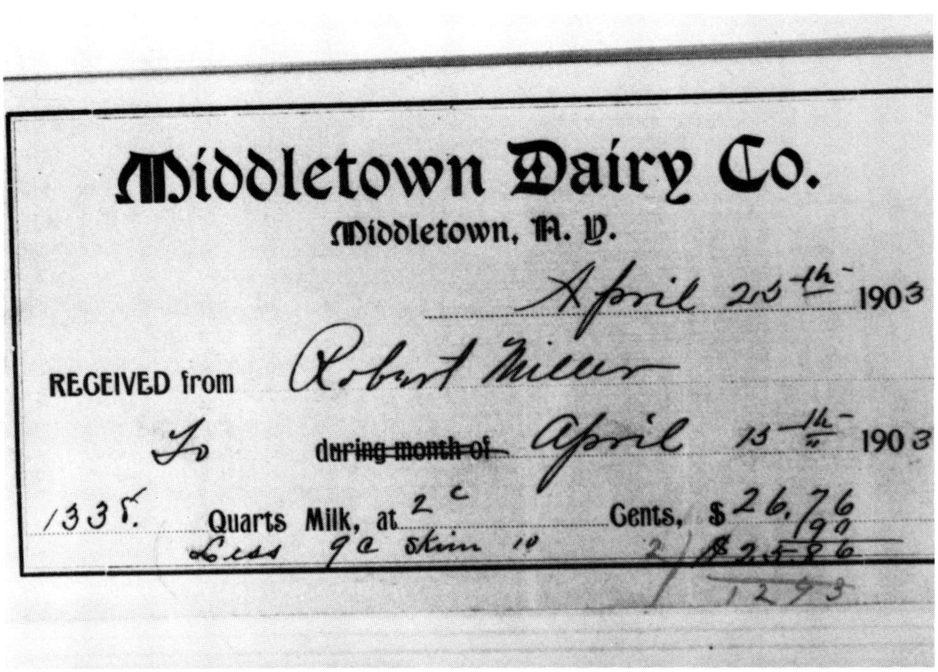

Middletown Dairy Co.

Middletown, N. Y.

April 20th 1903

RECEIVED from Robert Miller

To ~~during month of~~ April 15th 1903

1335. Quarts Milk, at 2c Cents, $26.76

Less 9c skim 10 .90

$25.86

Copy of Post Card-Bordens Creamery at Pepacton,N.Y.It stood next to the Railroad Station.

David Jacobson helped build the Breakstone's Creamery which stood farther up the Railroad Track.

Killed By Cruel Joke

Catskill Mountain News-May 25,1989-From the News Yellowed Pages-75 years ago-May 29,1914-"Stephen Wilson, 51,of Andes,an employer of Maynard Shaver,died after suffering for eleven months from drinking milk with muriatic acid. He had delivered a load of milk to the Pepacton Creamery and was given a glass.Upon drinking the milk he said it tasted funny,but was assured it was alright. Soon after he was taken violently ill and was never able to eat solid foods again."

Pepacton Creamery Crew

Left to Right - Sam More, Harley Misner, Archie Allen, Fred Harrington and Dan Bryden.

Copy from Mabel and Clara Miller Collection.

Post Card View of the Shavertown Creamery located in back of the Post Office and Store. It was later known as Fletcher Hall and provided a place for Silent Movies and Grange meetings and dances. It was last used for the community's farewell party on New Years Eve when Shavertown went out of existence.

Dorothy Rice Collection.

Shavertown Creamery Crew from the Collection of Mabel and Clara Miller.
From left to right-Ed Russell, Fred Harrington, George Francisco, Lynn More, Archie Allen, Charley Reside with Apron, Earl Barnes, the dog in front is Lynn More's dog Budd. Photo was taken about 1933.

From the Catskill Mountain News 'Yellowed Pages', July 24, 1914, "The closing of the Shavertown Creamery Company's plant here and the failure to pay for a months milk is a great misfortune to the farmers in this section. The creamery had a large patronage and thousand of dollars is probably lost by the farmers. No reason has been assigned for the failure to pay and people are at a loss as to what is the trouble."

Another news clipping from the Catskill Mountain News 'Yellowed Pages', August 7, 1914, REWARD OFFERED $500 reward: To anyone who will prove that I ever had any connection, directly or indirectly, with the Shavertown Creamery Company. Louis Kadams, Owner of Northfield, Shavertown and Dunraven Creameries."

From the Catskill Mountain News 'Yellowed' pages, July 14, 1939, "The dried milk plant and local creamery at Shavertown, which was once operated by the East Branch Creamery Company and employed several local men, is in a demolished condition since a crew of workmen started to tear the buildings down." 46

Air View of the Hood Place - The Cat Hollow Road joins the back river road at left of the small barn - the cattle barn and house are on the right of the small barn. The Cat Hollow brook runs near the Cat Hollow road and on across the field in the tree covered area to the river which you see at the right of the photo. In the background the school house is at left and the Charles Williams farm is on the right. The railroad track is the straight line which runs to the right of the back river road.

The Hood place was formerly owned by O.H. Scoville.

Photo of Hood place and post card view of Holstein cow and calf come from the collection of Mabel and Clara Miller.

Copy of a Post Card.

"Champion Holstein Cow of Delaware County. Owned by O.H. Scoville, Downsville, N.Y."

Jim Miller's Grandfather had a farm by Sugarloaf Road. Rob Miller (Jim Miller's Father) had a farm on Cat Hollow Road "Brock Farm". Then he moved to a Holiday Brook Farm (Tom Brown Farm). The move was up Little Spring Brook Road with a pair of gray horses "Maude and Tom". Then Rob and Jim Miller moved to the next farm up Holiday Brook.
The photos above and below are of the last farm owned by Jim Miller on Holiday Brook.

Both photos come from the Mabel and Clara Miller Collection.

From left to right are Oscar Felton, Art Hines, John Shaver and Jim Miller at the Miller farm on Holiday Brook.

Photo is from the collection of Mabel and Clara Miller.

Jim Miller is shown in his Model T car.

Mabel and Clara Miller Collection.

Mabel and Clara Miller Collection.

In front, at left - Millie Brown, Ila Tiffany, Maude Terry, Ellen Tiffany, Flossie VanSteenburg, Mabel Miller and Melisa VanSteenburg. Second row, at left - Willis Terry, Marshall Tiffany, Henry Tiffany, Charlie Tiffany and Claude Tiffany.
The above photo was taken on the Tiffany farm (by an old well). Later this farm became part of the David Jacobson farm on Holiday Brook. This land has since been subdivided and provides a weekend or seasonal retreat for people from the metropolitan area.

Photo at right is the house that was on the farm of Jack Bouw along Route 30 at the intersection of the Cole's Clove Road.
It was owned by Jack Bouw at the time the Pepacton Reservoir was built. In former years it had been the Sprague farm and before their time it had been the Scudder farm.
Clifford Sprague Collection.

Photo above-Auction at Jack Bouw farm near Pepacton,N.Y.as he prepares to leave the area due to the construction of the Pepacton Reservoir.

At right-Jack Bouw on the tractor.

Both photos come from the collection of Rev. Jacob Bouw.

This farm was located along the back river road between Pepacton and Shavertown. At one time it was owned and operated by Scott and Mary Barnhart, but in later years it was owned and operated by Don and Frances Bramley until it was taken for the construction of the Pepacton Reservoir.

Photo comes from the collection of Don and Frances Bramley.

The buildings in the photo at the left were owned by Ina Seath at the time of the construction of the Pepacton Reservoir. The building at the right in the photo was the Richard Woolsey Store - Shavertown's first merchant.

Betty Berke Collection.

Photo from Betty Berke
Collection.

The house in the photo above was located near the covered bridge which crossed the Beech Hill brook. It is marked G. E. Shaver Hotel in the 1869 Atlas of Delaware County, N.Y. by F. W. Beers. It was the original house on the Terry farm which was purchased by David C. Hoag in 1898. In his diary dated March 2, 1898, David writes, "We moved from Perch Lake to Shavertown."

This house had a very large fireplace which was located in the center of the house and one night a spark flew out from the fireplace and the house was totally destroyed by fire.

In a letter, dated March 1, 1901, David's sister-in-law, Mary Eleanor Fenton (later Reynolds), writes, "Dave has a house nearly ready to move in. He has moved Ira Miner's old house, (Ira gave it to him) and set it about half way between the barn and where the old house stood. They moved it today. George and I went down. They put runners under it and hitched six teams to it and started. It was quite a sight to see them drawing that house up through the fields. When they got opposite Finkley we went out and followed the crowd. They must have been an hour going from there to their destination. Crossing the binnacles and such things took longer than all the rest. George was the only one hurt. Someway he got thrown over one of the runners. Every one said he was lucky to come out alive, as it is he has quite a lame back. Will Coulter is bathing it for him now."

Later on, additions were built on to this house and it eventually became an eleven room house until it was destroyed for the Pepacton Reservoir.

A HUSTLING FARMER.

One of the Few Who Makes Tilling the Soil Pay--How He Does It.

[Special Correspondence.]

David C. Hoag is moving from Perch Lake to the Terry farm at Shavertown, which he purchased last fall. Mr. Hoag is one of the few farmers who has made a success at farming within the last ten years. When he bought the Washburn farm at Perch Lake ten years ago, it would not keep over six cows and team, and some people said the farm was at its best, for it was so rough that it could not be plowed; but David was young and strong and managed to turn over a patch here and there, and top dress the rest, so that in a few years he could easily keep twenty cows. He was busy getting out lumber to build a summer house at the lake to accommodate picnic parties, etc., and when Mr. Stanley, of Brooklyn, came along and offered him $1,000 more for the farm than he had paid he concluded to sell, and has bought and moved on to one of the best farms in the town of Andes, and with the help of his estimable wife, will strive tc make it a pleasant and permanent home.

Copy of newspaper clipping describes David C. Hoag and his farming enterprise as he moved from Perch Lake to the Terry farm which was located one mile below Shavertown next to the Lower Shavertown covered bridge crossing over the Beech Hill brook.

Below is a view of the former Terry farm which was purchased by David C. and Christina Hoag one mile below Shavertown and in later years it was owned by George F. and Letha Hoag until the Pepacton Reservoir was built.

This photo was taken just before the Pepacton Reservoir was built.

David C. Hoag above with the crow on his shoulder.

The following from the previously mentioned letter that told how another house was moved to replace David Hoag's house that had burned also tells of the social life during the year of 1901.

"Dear friend Mary,

"Your letter received Tuesday, was very glad to hear from you. Everything is as quiet as usual. The social so long promised has not come off yet.

"Will Coulter came here Tuesday and is here yet cutting wood. He will finish tomorrow. Frank was at a dance last night at Amos Parkers. They had eight numbers. That is the kind of dances he has attended all winter." ...

"The pattern was down to Tina's, but I got it today and will cut one like it and send it next week, and, you may keep it or give it to your Aunt Etta. I don't care which. You know she wanted it but I never got it up to her. I had to cut my skirt smaller than the pattern. You know Tina cut it for herself, and she always wants plenty of room in her clothes.

"Your Uncle Caleb has hired Maggie Mullinex for until after haying paying her between $4 and $5 per week.

"I am going to Andes tomorrow to get material for tacks. Then one day we are going to have a tacking. Wills people are coming over, and then maybe we can coax Lee to come over at night and we will have a dance. Wills are down to Toms tonight. We stopped when we came up tonight and left a can of milk for Stewarts baby. They keep it down there about half the time this winter.

"It would have been quite impossible for me to have gone to town today. We are expecting company tomorrow night and I want to be away a few days next week, so you see if you cannot come here neither can I go there.

"I hope you will have a good place and will write often, as I do not expect to see you before the Shavertown Fair."

Your devoted friend,

M. Eleanor Fenton"

Moving Tom Brown

by Bob Jacobson

It was a cold winter day in 1921 when Elmer Paul of Berry Brook and Jim Miller and David Jacobson of Holiday Brook each harnessed a pair of horses and hooked them to wagons for a trip to Walton. Elmer Paul had a pair of black stallions. Jim Miller's team went by the name of Rob and Roy and David Jacobson had a pair of mustangs named Prince and Kit. They were going to move Tom Brown's personal belongings from the village of Walton to a farm on Holiday Brook which would be a distance of twenty miles.

When they reached Walton the teams were kept at the Austin Sales Stable and the teamsters stayed at the Brown residence for the night. It was 5:00 P.M. the next day when the wagons were finally loaded and ready to start on the long haul up Bear Spring Mountain, presently New York State Route 206. The horses had to be rested many times before they reached the top of the mountain. When they started down the mountain it was a different matter! The heavy loads pushed and it was very cold and the teams were eager to go. Downsville was soon reached at a gocd fast pace!

The horses were unhitched at Downsville and put in the Ed Smith Hotel barn where they were fed and watered. The teamsters ate at the Oakley Shaw and Vernon Shaver Restaurant. After this welcome break, the horses were rehitched to the wagons and they were on their way up Route 30 to Pepacton. Once they had gone across the covered bridge at Pepacton, it was another mile up Holiday Brook to the place of Tom

Brown's destination. They reached this destination at 11:00 P.M. David put his loaded wagon on the second story of the dairy barn and since it was a dry night the other two loads were left outside. Jim lived on the next farm above Tom's place and David lived on the next place above Jim Miller's farm. Elmer Paul was not quite so fortunate as he had another 4½ miles to go beyond David's place. Their days work had reached far into the night!

Corn Harvest at Learia Sprague's farm. The 'unclaimed' are buried in the Pepacton Cemetery on New York City land which was a part of this farm along the Back Reservoir Road.
Photo is from Dick and Dora Sprague Collection.

Copy of Post Card - From Dick & Dora Sprague collection.

From a News-Tintype of Yesteryear-clipping,

"Gee! Haw! The late Levi Sprague of Shavertown holds the reins of Bright & George, oxen hitched up one summer day for a tour through the countryside. Mr. Sprague was the father of Ichabod Sprague of Walton, & in the back seat are Melissa & Jerry Sprague, sister & brother of Ichabod Sprague & his mother, Mrs. Alda Sprague. Seated by the driver is Edward Bresee. All of the quintet are now dead. A Sunday afternoon drive in those days would often result in an invitation & dinner from a cordial neighbor. Ichabod Sprague says these oxen would trot as fast as any horse. Bright, at left, blinks & drools at the cameraman."

These oxen were seen in the Shavertown area prior to about 1915. Notice they were not using an oxen yoke but were driven as horses with a harness.

Holiday Brook Property Owners Jubilant Over Decision Confirming First Business Damage Awards in Pepacton

Left to right, Alonzo Van Bramer of Berry Brook, Isaac Jacobson, Attorney Herman Gottfried, J. B. Bouw, John H. Shaver, David Jacobson examine decision of Appellate division, which confirms awards to these property owners for business damages caused by construction of the Pepacton reservoir. These are the first business damage awards made in Pepacton-Shavertown area. These five, plus Arthur and Gracia Hav·ns and Mrs. Anna Berg, will receive a total of $23,500. No real property is involved.

Copy of a newspaper clipping
from David Jacobson Collection.

Tourist Homes

It sometimes seems like the New Year should have started in the Spring instead of January 1. April 1 seems more appropriate. The summer birds arrive and fill the air with their songs and start to build their nests and the peepers come out on warm spring nights with their cheerful songs. I have memories of traveling from Shavertown to Margaretville along Route 30 during the 1930's and 1940's before the Pepacton Reservoir was built and the songs of the peepers echoed all the way. It was time for fishing and other summer activities would soon commence.

There were many tourist homes in Delaware County. During the era of World War II when gas was rationed, we developed a pattern of going to Margaretville on Saturday night for groceries at the A & P Store and other shopping at George Harris' Department Store and then we would go to the movies at the Galli-Curci Theatre.

As we traveled up the valley in the spring along with the tune of the peepers one could not help but think that summer boarders would soon arrive and walk along Route 30 admiring the beautiful country scenery. An elderly couple that often boarded at Carrie Elwood's Hotel in Shavertown would soon be standing on the iron bridge at Shavertown to fish in the East Branch of the Delaware River, and up above Shavertown there was the Norris Homestead. Next to the Jacksonburg Bridge was the Walter Hanselmann farm and up above Arena the Toth farm would also be filled with boarders. On the way home the area of Route 30 in front of Stan's Tavern above Arena would be lined with many cars. Stan's Tavern was a popular spot for Saturday night dances.

Copies of the following advertisement were found among the Hoag Collection.

WHEN one takes the route of much travel and journeys up the grand, majestic Hudson, he cannot but marvel at the wondrous river scenery stretching away on either hand until fading on the east in the vicinity of New England and on the west to the Catskills, haunts of Rip VanWinkle.

The Catskills, wondrous branch of the great Appalachian range, piled one against the other, austere in their height and glory. Every change of season or weather produces some new hue and mystical form noticeable at any time, but in the evening magnificent after the sun has faded far in the west and shades of darkness draw down on the close of a rural day.

In these hills of glory lies a little valley, parted by a stream silver in the sunshine, flowing just enough to play a melody that once heard is never forgotten. Near this river lie fertile meadows running out until they meet the woodland and pastures.

Among such surroundings is the Merry Meadows Farm, in the valley of the East Branch of the Delaware.

We trust for a continuance of your favors and hope to meet the best approval with our produce.

Mabel and Clara Miller
Collection.

The Downs House at Downsville, N.Y. was situated across the street from the present Downsville Diner where Sue Ellen's Country Store presently stands.

The last four men at the right of the photo above are identified as Pete Miller, Frank Radeker, Bill Taylor and Myers Jacobson. Upstairs on the porch are Harry Wright and Mrs. DeWeir. Mr. and Mrs DeWeir ran the hotel. Mullinex Clothing Store used to be in the higher part of the building.

The Norris Homestead

The Norris Homestead is located at Shavertown, New York, in the heart of the Catskill Mountains on the East Branch of the Delaware River. It is a large modern farm house located one-half mile from the village. The big feature of the Norris Homestead is its wide shady lawn and the Norris Eddy, noted for its fine bass and other fish. Also for boating and bathing. Our own garden supplies fresh vegetables and fruit in season. We also supply our own milk, cream, chickens and eggs from the farm. Home cooking and baking.

To reach Shavertown from New York by rail take

West Shore train at Weehawken for Arkville then change to D. & N. for Shavertown. Or take Ontario & Western at Weehawken for East Branch then change to D. & N. for Shavertown. The farm is one-half mile from Shavertown and guests will be met at depot.

Shavertown is a lively country village with three churches, Presbyterian, Advent and Catholic, three stores, two barber shops, some 75 residences and many advantages.

Rates $14-$16-$18 per week for rooms and meals. For further particulars write

Norris Homestead

Thomas E. Davis, Proprietor **Shavertown, N. Y.**

The description of The Norris Homestead at left is from the collection of Grant March.

The house at right is the house on the Norris Homestead. The property was owned by Thomas Davis until the construction of the Pepacton Reservoir. Dorothy Rice Collection.

Photo at left
is of the barn on
the Norris Home-
stead.
Dorothy Rice
Collection.

Catskill Mountain News, Margaretville, N.Y. Thursday, March 29, 1990--Page 3B

Photo courtesy Kathleen Dulaff

$24 A WEEKEND -- That's what it once cost to stay at the Toth Farm in Arena, a popular summer boarding house for 50 years, operated by Mary and Lazlo Toth. Guests enjoyed Mary's Hungarian cooking, Friday night dances, bowling and movies, as well as swimming and picnicking by the East Branch of the Delaware which ultimately claimed the farm when it was dammed to create the Pepacton Reservoir.

The Toth farm was a very busy place during the summer. Many city people came there to spend their summer vacations. It was located along Route 30 above Arena. Many people will remember the number of boarders that often walked along Route 30 admiring the country atmosphere.

Sidorowicz Tourist Farm,
Arena, owned by Stanley Sidorowicz was situated along Route 30.
Chip Sidorowicz collection.

The above post card view of Fenton's at Perch Lake where at one time many city boarders used to visit, comes from the Chip Sidorowicz collection.

CRYSTAL SPRING FARM

SHAVERTOWN, N. Y.

IN THE CATSKILLS MRS. JENNIE B. LIDDLE, HOSTESS

CRYSTAL SPRING FARM, ON THE TREMPERSKILL

Splendid trout fishing free to guests.

Board by the day or week. Rates reasonable.

Crystal Spring Farm is four miles from Andes and four miles from Shavertown, N. Y.

Phone No. 6R211

MRS. JENNIE B. LIDDLE, Shavertown, N. Y.

This place no longer serves as a tourist home but it still stands along Delaware County Route 1 and it is presently owned by Mrs. Oscar Babst.

The above photo was taken in front of the Pepacton Hotel, in the back seat of the car is Alice and Guy Williams who ran the hotel.

The photo above comes from the Mabel and Clara Miller collection.

Photo at left is a view of the foot bridge which crossed the East Branch of the Delaware River above the Walter Hanselmann farm and below Arena. The Jacksonburg School house was located by this bridge.

The Jacksonburg Bridge photo comes from the collection of Chip Sidorowicz.

Fishing

The following is written in the "History of Delaware County" by W.W. Munsell & Co., "An important source of food supply in the earliest time seems to have been the shad fishery in the river. In the investigation of the title to the land between the branches of the Delaware river in 1785, a number of affidavits were submitted .... Joshua Pine, Jr., aged twenty-four years, being duly sworn, deposith and saith that his father having purchased lands of Jno. Walton's patent, on the west side of Cockquago branch of the Delaware river, he the deponent, went with his father to settle there in the month of May, in the year 1785; that some time in the month of June in the same year the deponent went down the Pawpacton, or east branch of the Delaware river, with a canoe from the settlement at Pawpacton to Schehawken, and thence up the west branch to Walton's patent;...that shad did not run up, the west branch so high as Walton's patent, to the knowledge of any of the settlers, but that the shad came up to about cookhurse (the 'cookhouse' - Deposit)' and also that the people of Pawpacton told the deponent that they had caught thirteen hundred shad the year before, at one haul, in Pawpacton river; that the deponent never heard of any such quantity being caught in the west branch. Another deponent, Petrus Dumond, declares "that during the time he lived on the East branch of Delaware, near Paghkatacan (beginning in 1763), he frequently fished for shad below Papacunk (Pepacton) during their season, as also above the mouth of Beaverkill or Whelenaughwemack, when he caught large quantities of shad. This deponent remembers the time when the white people settled at Papacunk caught as many shad at one fishing about three miles below their settlement, as served the whole of their families for that season, as this deponent was informed.[47]

The fishing along the East Branch of the Delaware River has attracted many people over the years.

The aforementioned elderly couple who often fished from the iron bridge at Shavertown was a familiar sight for many years. There were also some other people who tried their luck from the bridge from time to time.

Corrine Griffith, the star of the aforementioned film "Single Track" which was filmed along the D & N Railroad, the film makers headquarters being located at the Pocantico Inn at Margaretville, enjoyed the trout fishing along the East Branch of the Delaware River.[48]

A. J. McClane, Editor of Field and Stream and author of several books about fishing spent summers in Margaretville when he was growing up and fished around Margaretville and Shavertown. He claimed to have learned about fishing from a master fly fisherman, Dan Todd, Station-master at the Ulster and Delaware Railroad station at Arkville.

Charles S. Woodruff writes, "I often recall the old happy days and the loyal friends residing in Margaretville, in the horse and buggy days. After my aunt, Miss Sarah Gleason, married Jefferson Allison of Dunraven, as a boy of twenty I spent one week each spring with the Allison family, fishing Dunraven, Mill and Dry Brooks with Dr. Reed, Dr. Tedford, J. H. Hitt and others. In 1889 I attended Jefferson Allison's funeral at Margaretville." ..... "The two happiest trips I ever made were going down the Delaware River from Margaretville with Dr. Reed, Dr. Tedford, J. H. Hitt, Amos Allison, Will Hubbell and others. Uncle Rastus Clute was our steersman and he knew every rift and stone from Margaretville to Trenton, N.J. The sweetest music I ever heard was

Bill Laidlaw playing "My Old Kentucky Home" on his cornet. I can still hear the echoes along the Delaware." ...[49]

As one travels around the Pepacton Reservoir now there is very little evidence of any past activities and only a very few landmarks remain.

Many who once lived there go back occasionally and as the water recedes sometimes a little piece of road and a few concrete bridges rise out of the water and occasionally if one is really 'lucky' they might find a piece of stonework or steps that were once connected to a building!

Up at Dunraven in a buffer area of New York City land there remains an old root cellar which once provided storage for vegetables during the winter months on the former Bruce Kelly farm.

It is hard to believe such a 'silent' area could have such activities as rafts going down the river in the Spring to lumber yards in Philadelphia, tourist homes filled with city boarders, roosters crowing, a train whistle, milk cans rattling, milk trucks and school buses going up and down the valley and farmers cutting hay and corn and growing cauliflower and other vegetables which would be shipped to New York City. It would have been a great tourist meca if the valley were still here with all the beautiful farms along the river, the covered bridges, the old iron bridges, the villages with the old store buildings and tree lined streets and the old school houses scattered around the area.

The area now appears like it may have appeared to the first white man only the river is larger and the trees are smaller! It no longer provides a hunting ground for Indians but the reservoir does attract many fishermen.

Bibliography

1. Delaware County History of the Century 1797-1897 Centennial Celebration June 9 & 10, 1897" Edited by David Murray LL.D, William Clark Publisher, Delhi, N.Y. 1898

2. "History of Margaretville and the Surrounding Area" by Ethel Bussy Coulter 1960

3. "Chapters in the History of Delaware County N.Y." by John D. Monroe, Honorary President and Historian, Delaware County Historical Ass'n. 1949

4. "History of Margaretville and the Surrounding Area" by Ethel Bussy Coulter 1960

5. "Chapters in the History of Delaware County N.Y." by John D. Monroe, Honorary President and Historian, Delaware County Historical Ass'n. 1949

6. "The Old Stone School House" by Lena O. Borden Tiffany

7. "Chapters in the History of Delaware County N.Y." by John D. Monroe, Honorary President and Historian, Delaware County Historical Ass'n. 1949

8. Letter from Foster Beach

9. Letters from Robert Hoag

10. Hoag family memorabilia

11. "Delaware & Northern and the towns it Served" by Gertrude Fitch Horton

12. Mabel Miller, David Jacobson and Isaac Jacobson

13. "Damn Nuisance" by Harry D. Archer

14. "Delaware County N.Y. History of the Century 1797-1897 Centennial Celebration June 9 & 10, 1897" Edited by David Murray LL.D. William Clark Publisher, Delhi, N.Y. 1898

15. "Three Centuries in Delaware County N.Y." by John F. DeVine published by N.Y. Swiss Alps of Delaware County 1933

16. "History of Margaretville and the Surrounding Area" by Ethel Bussy Coulter 1960

17. Family memorabilia

18. "Echoes of the Past" by Margaret Turnbull

19. "Atlas of Delaware Co. N.Y." from actual surveys by and under the direction of F.W. Beers assisted by A.B. Prindle & others, published by F.W. Beers, A.D. Ellis & G.G. Soule 1869

20. "Echoes of the Past" by Margaret Turnbull

21. Mabel Miller, David Jacobson and Isaac Jacobson

22. Downsville News

23. Mabel Miller

24. "The Catskills from Wilderness to Woodstock" by Alf Evers

25. "Pioneers of the Beaverkill Valley" by Lena O.B. Tiffany

26. Hoag Notes

27. Edna Goyette

28. "History of Margaretville and the Surrounding Area" by Ethel Bussy Coulter 1960

29. Florence Russell Huggins

30. Hoag-Jacobson Collection

31. David Jacobson - Holiday Brook School Record

32. Walton Reporter Jan. 26, 1940

33. Clara Miller Collection

34. David Jacobson, Mabel Miller and Clara Miller

35. "History of Margaretville and the Surrounding Area" by Ethel Bussy Coulter 1960

36. "Pepacton" by John Burroughs 1881

37. Florence Russell Huggins

38. Catskill Mountain News "Yellowed Pages"

39 News Clipping from Olive Wilson Collection

40. Hoag Collection

41. Andes Recorder

42. "Three Centuries in Delaware County" by John F. DeVine 1933

43. "History of Margaretville and the Surrounding Area", by Ethel Bussy Coulter 1960

44. David Jacobson

45. "Three Centuries in Delaware County", by John F. DeVine 1933

46. Catskill Mountain News "Yellowed Pages"

47. "1797 History of Delaware County, N.Y." W.W. Munsell & Co. 1880

48. "Damn Nuisance", by Harry D. Archer

49. Catskill Mountain News 'Yellowed Pages' June 9, 1939

Post Card View of Pepacton.

Clifford Sprague Collection.

